MODERN BIOTECHNOLOGY IN POSTMODERN TIMES?
A REFLECTION ON EUROPEAN POLICIES AND HUMAN AGENCY

Modern Biotechnology in Postmodern Times?

A Reflection on European Policies and Human Agency

by

Lars Reuter

Centre for Bioethics,
University of Aarhus, Denmark

SPRINGER-SCIENCE+BUSINESS MEDIA, B.V.

A C.I.P. Catalogue record for this book is available from the Library of Congress.

ISBN 978-94-010-3768-6   ISBN 978-94-007-1015-3 (eBook)
DOI 10.1007/978-94-007-1015-3

Printed on acid-free paper

Printed in the Netherlands.

TIBI OMNIA MANUS

TABLE OF CONTENTS

PREFACE   ix

ACKNOWLEDGEMENTS   xi

CHAPTER 1: BIOTECHNOLOGY IN MODERN TIMES 1
1.   The concept of modernity 3
2.   The concept of biotechnology 10
2.1.   Definitions 10
2.2.   Applications 17

CHAPTER 2: COUNCIL OF EUROPE POLICIES ON BIOTECHNOLOGY . . 25
1.   The Parliamentary Assembly 31
1.1.   The beginnings 32
1.2.   Developments of doubt and difficulty 33
1.3.   A new convention as a solution? 36
1.4.   Recent initiatives 39
1.5.   Results 45
2.   The Committee of Ministers 46
2.1.   Systematic beginnings 46
2.2.   The struggle for standards 48
2.3.   Results 54
3.   Council of Europe Conventions 54
3.1.   The European Convention on Human Rights 55
3.2.   The European Social Charter 56
3.3.   The European Convention on Human Rights and Biomedicine 57
3.4.   The Protocol on Human Cloning 61
3.5.   The Protocol on Transplantation 62
4.   Conclusion 63

CHAPTER 3: BIOTECHNOLOGY AND HUMAN AGENCY 69
1.   The human act 72
2.   Technique and technology 77

3.   Modern concepts of agency and identity 82
4 .   Is 'biotechnology' merely a name? 111

CHAPTER 4: BIOTECHNOLOGY BEYOND MODERNITY? 125
1.   A postmodern sketch of the human self 127
2.   Technology as bluff 135
2.1.   Human mastery 137
2.2.   The ambivalence of new techniques 140
2.3.   The dominating role of technology 145
2.4.   Conclusion 151
3.   The politics of contemporary biotechnology 154

APPENDIX 161

European Convention on Human Rights and Biomedicine
ETS 164 (Full text) with list of declarations 162

Additional Protocol to the Convention on Human Rights and
Biomedicine on the Prohibition of Cloning Human Beings
ETS 168 (Full text) with list of declarations 185

Additional Protocol to the Convention on Human Rights and
Biomedicine on Transplantation of Organs and Tissues of
Human Origin. ETS186. (Full text) 191

REFERENCES 207

INDEX 225

PREFACE

Dealing with issues of biotechnology and bioethics in different European contexts, I began to realise that the various efforts undertaken to evaluate and regulate this field were inspired by a particular view of the human agent. For current biotechnology confides in the human agent's ability to instigate transformations of living and non-living objects into pursued states of being in various environments. Similar to other forms of technology, there is in this field a certain optimism with regards to the scope of human efforts and the effects they produce. Such an optimism is understandable for there has been technological progress, with which life at least in the Western World has become significantly simpler, safer and far more predictable. At the same time, there is some concern about this very development and public presentations of biotechnological products and procedures tend to fuel often heated debates about their ethical or legal legitimacy.

Against this background, I felt a need to investigate matters a little further, nourished by the demands of those who asked me for a concise introduction to the questions raised by current biotechnology and its regulation in Europe.

Already at this point, I need to stress, then, that by identifying an inherent dependence of biotechnology on modernity, I do not intend to mark the latter as an inherent evil. There is, thus, no underlying normativity in my analysis with regards to this concept, even though the analysis might dissect in ways that will be perceived as highly critical. Simply put, I find it important to contemplate upon the actual structures and potential functions of biotechnology in our societies in order to allow for a somewhat more balanced discussion in this regard. Therefore, I would very much like this book to stimulate such common reflection, which I hope will be possible in spite of its clearly European origin.

This book is, thus, combining two main foci: first of all, it presents critically the deliberations of these issues in the Council of Europe, which has been deliberating the regulation of biotechnology for decades. As a major forum of intergovernmental policy-making, it clearly demonstrates the richness as well as the lacunae of such a broad European enterprise. Since it seemed quite difficult to find a concise presentation of the positions on biotechnology defined by the two main bodies of the Council of Europe, namely the Committee of Ministers and the Parliamentary Assembly, I decided to produce a comparison as it is presented here. Proceeding chronologically, I wish to show how the deliberations on biotechnology are closely intertwined with the assessment of benefits and dangers emerging from research and product development in this field. Beyond giving an impression of how European policies on biotechnology evolve, my reflections also attempt to reveal the underlying struggles to define values, standards, and principles. In the course of this investigation, it will become clear that biotechnology as well as the policies drafted to regulate it rest upon the idea of strong human agency. The human subject is seen as the natural point of revolution, then, by which organisms and other bio-chemical entities are used and changed in accordance with the will of the agent. This agency presupposes that the human individual is able to fathom the reality of e.g. genes and to change it in accordance with anticipated ideals. Hence, the human agent is understood as an examiner and creator, i.e. as a subject dealing with objects, which is a fairly modern way of thinking, of course.

If modernity no longer is the prominent paradigm for societal interaction, however, then the modern view of biotechnology suddenly seems somewhat anachronistic. This is the second line of thought I shall follow. For this purpose, I analyse the modern understanding of the human act and the agency it presupposes in order to see whether it still can be applied, in particular in view of the postmodern positions presented subsequently. This analysis will draw to its close by relating the thinking of one of the prominent critics of modern technology, namely Jacques Ellul, to the development of biotechnology in order to determine whether his almost archetypical form of criticism pertains to our field at all.

I realise that the composition of this book may seem less stringent than could be expected. In particular, there will be a significant, but fully intended change of mode from chapter two to three. At this point, the reflections on policy will feature as sources of inspiration for reflections on human agency and the acts stemming from it in technology and beyond. Metaphorically put, it is journey into the heart of matters, beginning with the broader context, then displaying the current state of regulation, taking the Council of Europe as *pars pro toto*, followed by a dissection of the concepts that I identify as vital for biotechnology and its regulation. The main reason for choosing such an approach is not to disturb the reader, of course, but to combine political, legal, ethical, philosophical and theological reflections on current biotechnology and its human agent, simply because biotechnology itself appears as formed by a variety of interests and initiatives that surpass the traditional concepts of academic disciplines. In this, my contribution is, perhaps, a hallmark of postmodernity, granting that reflections on issues that so deeply affect society no longer can be presented along lines that would pretend a stability of the object we wish to analyse.

Let us begin, then, with defining this misty field of biotechnology and the time in which it occurred.

ACKNOWLEDGEMENTS

The writing of this book was substantially supported by a University of Aarhus research foundation grant and a Hoover Foundation Fellowship, both of which allowed for my staying at the Centre for Biomedical Ethics and Law, KU Leuven, during spring term 2002. I owe gratitude to the staff at the centre for providing me with inspiration, challenges, and utter kindness.

Ms Annette Larsen, secretary at the Centre for Bioethics, University of Aarhus, contributed significantly to the preparation of the camera-ready copy of this manuscript, for which I thank her very much.

A final word of thanks is due to all those who endured me and my ideas during the writing process.

CHAPTER 1

BIOTECHNOLOGY IN MODERN TIMES

'Medieval biotechnology' would to most be an oxymoron.[1] This is hardly astonishing, for the common understanding of this human activity and its means and ends seems to be as deeply embedded in modernity as the very concept of technology itself. Is speaking of biotechnology in modern times merely a truism, then, rendering it superfluous to investigate matters further, since there simply is no other time in which biotechnology has been, presupposing, of course, that these days still are *modern* times? Perhaps. Yet, if the present is no longer 'modern', but e.g. postmodern or not modern at all, biotechnology might have lost its grounding, if not its entire *raison d'être*. This chapter will, therefore, begin with defining the concept of modernity, after which we turn to biotechnology and its *situation*, the latter understood in the Sartrean terms of the subject merely passing by the object without already instrumentalising them in the same act.[2] I intend to do precisely this, namely to see biotechnology simply as it *is*, namely as being there, while the questions of necessity or possibility relating to its being will be dealt with later in the book.

I take biotechnology to be a true project of modernity, then. This is the first part of my thesis, based upon my view of the modern project and its maker. The second is

---

[1]The nature of the oxymoron is, of course, to create ambiguity by virtue of its inherent contradiction, which is uncalled for in our present context: "L'oxymoron est ambigu, il n'est pas pour autant en équilibre, car un des deux termes semble vouloir rattraper l'autre. Ce n'est pas une ambivalence comme l'amour et la haine qui peuvent se côtoyer de manière symétrique chez des amants. (...) Non, le *"merveilleux malheur"* est lui de travers, il boite." De Brabandere (2002).

[2]Cf. Sartre (1956), 548f: "The situation can not [sic] be subjective, for it is neither the sum nor the unity of the impressions which things make on us. It is the *things themselves* and myself among things (...) that (...) simply are *there* as they are without the necessity or the possibility of being otherwise and that I am *there* among them (...) The situation is the subject illuminating things by his very surpassing, if you like."

1

that after the at least alleged crisis of modernity, this project now seems anachronistic, since the strong subject of modernity, on which biotechnology relies so heavily, has been weakened considerably, leading us to wonder whether that subject indeed can fulfill the role it has assigned itself. In this respect, it is worth remembering that the various forms of utopia envisioned from the late 1940s to the early 1970s also were modern projects that now have been abandoned, also quite literally.[3] Obviously, the notions of what modernity entails have changed, then, but how many changes would eventually void the concept?

Since the common endeavour of biotechnology does make an impression on 'us', i.e. society, and (parts of) our lives, its use has been discussed and regulated politically, e.g. by law. Furthermore, science is, like most of the academy, international in its scope, and so is biotechnology, especially because some of its foundations could only be established through the coordination of national initiatives, as in the Human Genome Project, begun in 1988[4] and preliminarily completed in 2000.[5] In Europe, biotechnology has been the object of coordinated funding as well as regulation. I shall attempt to illuminate its situation through the views on biotechnology offered in documents issued by the Council of Europe. This organisation forms a major European forum of debate and decisions on common economic, legal, and political policies with regards to biotechnology, reflecting varying levels of multilateral agreement reached on the European identity and scope of this field in these allegedly 'modern' days of ours.

---

[3] In several European countries, public housing projects involving pre-fabricated elements have proven to be quite unpopular and deserted buildings have been torn down, while others have been diminished, remodelled or at least renovated. A systematic reflection on this topic is found in: Prak (1985). With regards to wider bearings of social housing, cf. Danermaerk and Elander (1994). The social aspects of this type of housing are discussed in: Power (1997). She also gives a survey of the historical development in: Power (1993).

[4] Cole (1998), 55.

[5] The working draft was completed in 2000: "June 26, 2000 will be remembered as an important day in the history of science and milestone for humanity. The international Human Genome Project (HGP) Consortium, which is composed of scientists from China, France, Germany, Japan, United Kingdom and the United States of America, announced the completion of the human genome working draft at 18:00 p.m. (Beijing Time) that very day." Yang (2000), 63.

1. THE CONCEPT OF MODERNITY

The time of 'modernity' is often perceived as a period evolving from Europe through a series of decisive developments: the voyages to the continents beyond, the discovery of a solar system in which the earth is merely a planet, the emancipation of the human mind from its divine creator and, thus, also from the Christian religion proclaiming 'him', the separation between church and state with its concomitant shift of power to the latter, and the increase in the human domination of nature especially by the means of science, all of which contributing to, if not resulting in the state of anthropocentrism.

Henceforth, humans do not just understand the world in human terms, the inevitability of which is self-evident, they also shape it in accordance with their mind *solely* on its own account: "The modern concept of subjectivity rests solidly in the idea of freedom, so much that many consider autonomy *the* characteristic of the new age."[6] This expression of autonomy is firmly established in the enlightenment, severing the human agent from its mundane and metaphysical contexts. Consequently, the only point of reference left for human agency is humanity itself, notwithstanding its possible (self-)transcending abilities as assumed e.g. in social contract theories or any other position linking the (practical) reason of the individual with an assumed commonwealth of (theoretical) reason. Realising this autonomy, a modern understanding[7] of human agency excludes any element of *ontological* otherness from the human act, neither promising nor granting relief for the rolling of our stones.[8] The *human* act *is* nothing more than it is, apparently. This raises questions of the precise nature of the subject acting, especially concerning its potential for agency, to which we shall return in chapter 2, section 3.

It would be somewhat ignorant, of course, to regard modernity as a sharply defined period on a historical continuum. Historical periodisation is a tool of comprehension

---

[6]Dupré (1993), 120.

[7]This judgement does presuppose a particular view on modernity and its concomitant thinking. Depending on the how far this concept is extended, certain philosophical traditions or positions are not covered by it, e.g. a variety of phenomenological approaches that do seek to take ontological otherness into account.

[8]In this, the human subject would very much find itself in the situation of Camus' Sisyphus, who, as it is well-known, must be considered a happy man, because he has understood the inevitability of his situation. We might be condemned to freedom, but at least it is our own freedom. See Camus (1955).

rather than a description of reality, after all. It is, therefore, slightly more relevant to understand the term 'modernity' as a name for certain practices, attitudes, and relations marking human agents and their agency. In this sense, one might detect modern modes of being at various points in history, establishing forms of acute anachronisms, while maintaining inherent strains of synchrony in time. 'Modernity' as a concept may, thus, be parallelled with the way in which that of 'classicism' is employed: a specific notion of reading the world in art and architecture, re-emerging or rediscovered at various points in time since Greek-Roman classicism, e.g. at the turns of the 12th and 19th century.[9] A concept of time such as 'classicism' requires 'texts' in the sense of deliberate fixations of communication in writing, sculpturing, painting, or building construction, in this instance seeking its origin in the part of 'Antiquity' from which sources remain. Understanding the past as a sequel of distinct periods is a fallacy, because a period does not exist *per se*, but only to the extent people act in the ways they do, thus providing a folio for systematisation *a posteriori* with specific modes of actions, acts, and agency delicately dissected for this very purpose.

Granted, human actions are influenced by collective ideas such as societal norms and values, but the concept of a period first emerges when human actions by analysis are arranged to form precisely what one is looking for. Even periodisation faces the problem of human individuality, i.e. individual acts that do not conform with standardised behaviour, which sometimes is solved by classifying these acts of differing modes as "anachronistic". Logically speaking, they cannot be that, since any act is performed in its particular time. Inherently, no act is "anachronistic," but it may be deemed less in accordance with societal standards. In other words, the term is not descriptive, but normative.

One could define modernity in our present context, then, as a period in which modern modes of being are accepted as standard references for societal interaction and historical periodisation is employed to distinguish 'the past', here understood as 'history', from the present. The tentative openness of the last conjunction acknowledges that the present merely is future turning into past, evading an actual grasp of any kind, be it intellectual or physical, apart from the very elusive encounter

---

[9]The reason for distinguishing the concepts of 'emersion' and 'discovery' here is their difference in presupposition: is the historical period coined 'classicism' indeed an ontological idea emanating in time, the temporal thus shaped by the eternal, or is it merely a finite human vogue, having become a matter of fashion?

of its passing.[10] Moreover, this definition encompasses the co-existence of different contemporary periods in various parts of the world: if 'modernity' originally is a European concept, exported to other parts of the world by culture, war and trade in the course of roughly five centuries, it remains but one categorisation of human activity, complemented by all other forms of human civilisation. In view of the variety of concurring cultures, however, it is striking how the modern concept has secured its volatility compared to the alternatives: "There may be many forms of tribal society, many feudalisms, even many forms of early capitalism, but there is only one modernity and it is exemplified in our society, for good or ill."[11]

In competition with the other modes of societal identity and development, modernity has proven to be the most robust concept, formed and applied by individuals with very different backgrounds. Even if one were to argue that European colonialism (and its US-American heir)[12] had the force to mark, if not programme, the mobile members of societies anywhere, one would still need to account for the willingness with which people consume identical products and seek to conform with the standards of the Western hemisphere. This process is often called 'globalisation', which e.g. is characterised "as a new stage in the development of economies; while internationalisation points to an increasing border crossing of economic activities, globalisation implies the merging of national markets into one unbounded world

---

[10]This raises the question of the nature of time, which cannot be answered here. The volatility of the present have inspired concepts of overcoming its condition by eternalising it. In this way, S. Kierkegaard's notion of the moment in Philosophical Fragments may be seen as an attempt to provide a robust structure: "The temporal point of departure is a nothing, because in the same moment that I have known the truth from eternity without knowing it, in the same instant that moment is hidden in the eternal, assimilated into it in such a way that I, so to speak, still cannot find it even if I were to look for it, because there is no Here and There, but only an*ubique et nusquam*. If the situation is to be different, then the moment in time must have such decisive significance that for no moment will I be able to forget it, neither in time nor in eternity, because the eternal, previously nonexistent, came into existence in that moment." ibid., p.13 with a view over the underlying discussion in the footnote, where especially Hegel's position in *Differenz des Fichteschen und Schellingschen Systems* is relevant: "Das wahre Aufheben der Zeit ist zeitlose Gegenwart, d.i. Ewigkeit."

[11]Feenberg and Hannay 1995, 6.

[12]I realise that using the term 'colonialism' with regards to the USA is somewhat delicate. There is little doubt, however, that this country has political as well as economic interests also manifesting itself in forms of hegemony. A brief discussion of the hegemon status of the USA as opposed to other major world powers is found in: Huntington (1999).

market. It signals a growing integration and interpretation of economic activities on a world-wide scale."[13] The precise scope of this concept is debated. Göran Collste sees it primarily as awareness of increasing interdependence[14], while others detect limitations to the extent of seeing it as a myth, because globalisation has only taken place within certain regions and branches of industry.[15] It seems more appropriate, then, to regard this process as one of sometimes deliberate standardisation, at least as far as fashion[16], business[17], travel[18], music[19], IT[20] and the academy[21] is concerned, with

---

[13]Schienstock (2001), 51f.

[14]Collste (2001), 425, "the concept catches a wide spread impression that the world is shrinking and that, in an earlier unknown way, the lives of people living far apart are in many ways interdependent. There is a growing awareness that we are all sitting in the same boat."

[15]Schienstock (2001), 52, footnote 6: "Empirical data clearly show that up to now only within the regional economies of North America, Southeast Asia and Europe, and to a much lesser extent between these economies, has a process of economic integration and interpenetration started. Furthermore, globalisation has, as recent empirical studies show, progressed significantly only in a few industrial branches such as automobiles, chemical and consumer electronics as well as in some parts of the service sector such as banks and insurance, whereas other branches have been affected less by the globalisation process. (...) However, to conclude that globalisation is just a myth, underestimates the dynamics of economic development during the past decades and is definitely an exaggeration."

[16]Note e.g. the success of the Western models of the gentleman's suit (trousers, shirt, tie, jacket, belt) and the lady's dress (long with sleeves or two piece suit with blouse).

[17]E.g. the global stock market, now open 24 hours at least five days a week. Cf. also with the details of the business culture, such as the business card, the fairly standardised stationery etc.

[18]Take e.g. the full compliance with IATA (International Air Transportation Association) standards. It is remarkable that even peasants fleeing from tormented Somalia would be arriving on planes in Northern Europe.

[19]This ranges from Japanese children playing Mozart to the music channel MTV.

[20]The standardised internet access and setup, the almost complete hegemony of microsoft, the widespread use of branded search engines (e.g. 'Yahoo'), the duopoly of the Word Perfect and Word programmes etc. Collste (2001), 425f, describes the co-dependence of globalisation and IT as follows: "Information technology (IT), e.g. Internet and e-mail, has been a pre-requisite for the globalisation process. The global networks, or the Global Information Structure (GII) as it has been called, have made it easier and cheaper to communicate across the world to the extent that geographical distance in many instances has become irrelevant."

[21]The mere fact that I, although of Danish nationality, would write a book in English for a Dutch academic publisher. One could argue that, depending on the academic discipline, English has become lingua franca replacing Latin, German or French. Still, my point here is to show that regardless of your local heritage, you are adapting to the norms of the group to which you (would like to) belong and that group has now become a 'global player'.

remaining regional enclaves of local custom and folklore. As a deliberate enterprise, standardisation occurs especially in the technological sector, with the International Organization for Standardization (ISO) as "one main agent for technological globalisation."[22] In this context, standards can be defined as "documented agreements containing technical specifications or other precise criteria to be used as rules, guidelines or definitions."[23]

Since trade relies on the exchange of products, services and the tender representing it, a "main reason for global standardisation is to facilitate international trade (...) When international standards are widely used, suppliers can compete on many markets worldwide and customers have a wider choice of compatible offers. Another reason is that often one component, is used in different products."[24] Such forms of standardisation reflect particular interests, in casu reducing costs and increasing sales, which apparently meet the interests on the part of the consumers. Furthermore, they assume a freedom of human agency, expressed through the notion of choice, that is characteristic for modernity and its hallmark of technology. It is the technological development, then, that is accelerated by standardisation, resulting in possibilities and risks on a global scale.

The reactions on the attack on the World Trade Centre in New York, NY on 11th September 2001, have clearly demonstrated the power of this process, displaying an international public interest and resulting in a common response of military attack and tightened security measures affecting a remarkable number of travellers and insurance customers far beyond the borders of the USA.[25] Complete standardisation has not been achieved yet, in part because in every society, the relative number of those endorsing

---

[22]Collste (2001), 428.

[23]Ibid.

[24]Ibid. Collste uses the format of credit, phone and smart cards as an example, where the setting of standards such as an optimal thickness warrants international operability.

[25]The confiscation of nail scissors from one's hand luggage in preparation for a Danish domestic flight from e.g. Århus to Copenhagen is in this regard significant in several ways: first of all, it links the individual to an incident with which he is hardly connected directly; secondly, it transforms the psychology of traveling within one's own country: when crossing the lines of security check, I am moving into an international zone of tension and surveillance, even though the flight never leaves Danish territory and the domestic terminals are separated from the international; thirdly, it inflicts quite arbitrarily on the individual's property right on alleged utilitarian grounds: if it were a matter of security indeed, travelers should not be allowed to bring metal pointed umbrellas, laptops or glass bottles bought duty free, all of which would make for excellent instruments of attack.

it varies according to the degree in which this Western concept is known and embodied by the individual. I speak of embodiment in this respect, because the adaption of modernity is more than the superficial colour change of the chameleon: it is difficult to see how a so distinct way of dressing, eating, washing, communicating and exposing oneself to the arts will not affect the whole person, provided one does not regard human personhood as substantially immune to the influences of e.g. human acts.

Modernity is, thus, created by modern modes of being, since the most fundamental of these modes, namely that of periodisation, is the prerequisite for establishing a period in time called 'modernity'. If there were no 'middle ages', neither could there be a subsequent period, which by the very periodisation of time has erected itself as the epitome of human development: "The "modern" outlook is that there are periods which have distinctive and proper thoughts and attitudes. If this were not true, there would be no "modernity" and we would have to settle with novitas and these not as "modern times" but simply the dies novissimi."[26] In other words, "[p]eriodization is itself a modern strategy." [27]

Ironically, employing the concept of modernity may reveal an underlying attitude of conservatism, attempting to preserve what is fragile and perishable, namely the novelty of any moment and action, by placing it in a category that by definition is constantly up to date. Yet, repetitious actions, seeking to copy a template, will always be carried out in a new temporal frame and context, because any human action is embedded in time and space, parallelled with a finite, but very large number of actions taking place at the same time.

Whenever the past as such or its interactions are understood periodically, 'modernity' thereby proves itself. This statement may seem circular. It is only so, though, to the extent that e.g. J. St. Mill, when presenting the foundations of his theory of utilitarianism, with regards to its basic principle of utility could state that "all desirable things (...) are desirable either for the pleasure inherent in themselves, or as means to the promotion of pleasure and the prevention of pain."[28] In other words, the desirable is that which is desired, as Elijah Milgram convincingly shows, and Mill's

[26]Mark D. Holtz in an e-mail 14th April 2000, received at 1857 hrs GMT+1 at reuter@teologi.au.dk.

[27]Melehy (1997), 10.

[28]Mill (1985), 258.

argument, thus, an inference.[29] Likewise, the argument launched here is basically empirical, the conclusiveness of which depending on the acceptance of logic, however, very much similar to the definition of modernity in the OED (1989): "the quality or condition of being modern; modernness of character; something that is modern." Without being modern, no modernity, and without establishing 'modernity' through the act of periodisation, one could not identify something or someone as being modern. This logical truth is confirmed empirically, whenever someone uses the terms 'modern' or 'modernity' in the same sense as it is done here. Modernity is desired, then, but its desirability depends upon its very deliberate construction. The whole point of modernity is, then, "that it does *present* itself as a unity, as a period that may be narrated by a single subject[.]"[30]

There is also an inherent arbitrariness of historical periodisation, which is emphasised by the difficulty of firmly establishing the boundaries of such periods. With regards to 'modernity', the criteria have hitherto remained opaque: does it begin with the great voyages and expeditions, the renaissance, the reformation, absolutism, industrialisation, the enlightenment, the great revolutions, nationalism, capitalism, colonialism, parliamentarism, the ultimate suicide of the *anciens régimes* in the First World War, or fascism? And if it has ended, when did that happen: with the end of the Second World War, the movements of the sixties, culminating in the incidents of 1968, the 'oil crisis' in 1973, the alleged 'end of history' in the late eighties or the fall of the iron curtain just after it?

Since the aforementioned definition of modernity in the OED (1989) rests on the assumption that something can, indeed, be modern, the definition of that term is illuminating: "of or pertaining to the present and recent times, as distinguished from the *remote* past (my italics, LR); pertaining to or originating in the current age or period. In historical use commonly applied (in contradistinction to ancient and medieval) to the time subsequent of that time." It is clear, then, that the point made above is sustained: the concept of modernity is inherently dependent upon the periodisation of the time preceding it. With the earliest quote taken from T. Washington's translation of Nicholay's voyage, dating 1585, it is interesting to observe that according to this definition, the 16th century is not remote past, whereas the 14th century is. Following

---

[29]Millgram (2000), especially 289. It is classified as 'inference', since there is a movement from evidence to conclusion, with desirable equaling the desired or $p \rightarrow p$

[30]Melehy (1997), 9.

the argument of periodisation strictly and ad absurdum, any person living in 1585 would be closer to us than to his predecessors living in, say, 1485. At least linguistically, such a position would be difficult to defend for many, if not all areas in Europe. The case may be different in terms of human self-understanding, especially with regards to human agency and identity. In this regard, it is interesting that some would use the term 'modern' to describe the process of evolution as having led to modern mankind and relate this development to the size of the brain, evoking that this feature would be of special significance in terms of human identity.[31] The bearings of these questions will be discussed in chapter 2, section 3.

At present, it will suffice to maintain that biotechnology has emerged in a time that some wish to call modern.

2. THE CONCEPT OF BIOTECHNOLOGY

With the following two sections, the examination of this concept and its wider bearings begins. First of all, it is necessary to clarify the meaning of the words used to name this concept, i.e. to analyse typical types of definition employed in this field and relating it to the notion of bioethics. I shall also present my own definition, serving as basis for the subsequent study. Secondly, some applications of biotechnology will be presented in order to show the reality of this concept.

2.1. Definitions

As most terms of categorisation, the word 'biotechnology' is a homonym. This is already evident in the way the OED (1989) defines the word "1. the branch of technology concerned with the development and exploitation of machines in relation to the various needs of human beings; 2. The branch of technology concerned with modern forms of industrial production utilizing living organisms, especially micro-organisms, and their biological processes."

The first definition seems oddly broad, if not tautological, for what machine would not have been developed or exploited by humans for the very sake of - ultimately -

---

[31]"The outstanding feature in the evolution of modern man is the growth of the size of the brain, the greatest extent of this increase taking place in the cerebral cortex and its nuclei, and in the cerebellum." Duncker (2002), 57. See also ibid., 56: "...the development of the social and cultural beings of modern mankind."

serving the 'various needs of human beings'? The second definition is more to the point. Since biotechnology here is seen primarily as a *technological* instrument, it is shaped by advanced forms of industrial production, which I therefore take to exclude the forms of industrialisation dependent upon water, coal, iron, and wind. Biotechnology is in this sense a modern tool employed by humans for their own sake.

It is clear, then, that biotechnology and late modernity, i.e. the time *after* the industrial revolution, are inherently linked, since the former requires tools and knowledge first developed in the time of the latter.

The European Commission regards this transition as a "revolution (...) taking place in the knowledge base of life sciences and biotechnology, opening up new applications in health care, agriculture and food production, environmental protection, as well as new scientific discoveries. (...) The expansion of the knowledge base is accompanied by an unprecedented speed in transformation of frontier scientific inventions into practical use and products and thus also represents a potential for new wealth creation: old industries are being regenerated and new enterprises are emerging, offering the kind of skill-based jobs that sustain knowledge-based economies."[32]

Biotechnology does not merely comprise tools, then, together with the life sciences, it is depicted as the very foundation of new societies emerging from those based on old industries, which is expected to create work and wealth. The use of the term 'frontier'[33] is in this regard evoking 19th century images of ventures into territory unknown to Europeans, which, albeit already inhabited, would provide the fields for culturing the new world.[34]

For the commission, health care is one sector where biotechnological procedures

---

[32]Commission of the European Communities: COM (2002) 27, 3f.

[33]Phillippe Busquin, European Commissioner for research, uses this term somewhat enthusiastically, maintaining that biotechnology is "indeed the new frontier in science and technology and it can be a major driver of innovation and wealth creation in Europe.": Speech at the Belgian-Danish Forum for Innovation in Biotechnology. Brussel 29 May 2002. Unpublished manuscript, 3.

[34]According to Frederick Jackson Turner's thesis, "the existence of an area of free land, its continuous recession, and the advance of American settlement westward, explain American development (...) this frontier accounted for American democracy and character, and (...) at the end of the 19th century the continental frontier finally closed forever, with uncertain consequences for the American future." Quoted in: Faragher (1994), 1. Cf. also Klein (1997). Lee attempts to avoid the cowboy ethos by tracing "a critical genealogy of the narrative traditions through which historians, philosophers, anthropologists, and literary critics have understood the European occupation of Native America, and (...) how those understandings shaped and were shaped by changing conceptions of history." Ibid., 6.

will be applied. Since it is not as such evident, however, that these two areas are related, I need to clarify in what ways they indeed are. The proof quotations given in the OED (1989) may be helpful in this respect. In 1969, the Scientific Journal June 50 states: "Biotechnology is just as concerned with the provision of tools for medical research as with the development of equipment for medical service." In 1972, a periodical called Biotechnology and Bioengineering Symposion has appeared. And in 1985, I. J. Higgins in I.J. Higgens et al.: Biotechnology, i, 2, explain why the technology has grown popular: "It is...the discovery of genetic engineering techniques via recombinant DNA technology..., which is responsible for the current 'biotechnology boom."

At this first glance, biotechnology appears as an area of fundamental research, the role of which increased as its results in understanding genetic engineering proved to be of value especially for the medical sector. In 2001, others see its roots in food production, arriving at a similar conclusion, worth quoting at length:

> "Historically, biotechnology evolved as an artisanal skill rather than a science, exemplified in the manufacture of beers, wines, cheeses etc. where the (...) molecular mechanisms went unknown (...) The traditional biotechnology products have now been added to with antibiotics, vaccines, monoclonal antibodies and many others, the production of which has been optimised (...) in particular, recombinant DNA (rDNA) technology, which is now giving bioscientists a remarkable understanding and control over biological processes (...) It is most probable that this rDNA technology or genetic engineering (...) will be increasingly viewed as a branch of modern science which will have profound impacts on medicine, (...) the development of new biopharmaceutical drugs and vaccines for human and animal use, the modification of microorganisms, plants and farmed animals for improved and tailored food production and to increased opportunities for environmental remediation and protection."[35]

This statement collects the central elements of our investigation: when the subject did no longer consider it sufficient merely to produce, but necessary to fathom, say fermentation, biotechnology turned into science, marking a step from the subject as an agent of sheer making to that of analysis, with which the subject distances itself from its own manufacture and the results of it so that it may 'objectify' by way of dissection. The goal of this endeavour is to 'optimise', i.e. to better the imperfect and to gain true control over the ways of nature, which the knowledge of rDNA promises to grant, here

---

[35]Smith (2001), 3f.

expressed by the use of the term 'remarkable', signifying that the degree of understanding and control is now closer to the actualisation of its potential. Hence, the use of rDNA united the disparate efforts of examining and understanding the laws of microorganisms, finding cures for disease in medicine, improving the production of animals and plants, and dealing with the environmental mess we make. According to this understanding of rDNA, it pledges that humans will be able to finally live without fear after having overcome the contingencies of our existence. A little pejoratively put: in utopia[36], we will be protected by vaccines and drugs, produce precisely the food we want and make our damages to the environment vanish. This is indeed a remarkable enterprise, which "was foreseen in the mid 1960s and came to fruition in the early 1970s (...) From (...) relatively modest beginnings, techniques for manipulating and analysing both types of nucleic acid (DNA and RNA) have become remarkably powerful and sensitive (...) The advent of recombinant DNA technologies led to the realisation that DNA could be analysed to a resolution that was unimaginable only a few years before and consequently the genomes of almost any organism, prokaryote, archaea or eukaryote, could be manipulated to direct the synthesis of biological products that were normally *only* [my italics, LR] produced by other organisms."[37] Similar to agriculture, the human subject ventures to direct the genetic ways of nature, including the determined movements of "other organisms," thereby subjecting the inherent laws of necessity to human will. This extends the realm of human influence and may therefore increase the sense of autonomy.

The view of biotechnology as a traditional practice having been transformed by the new techniques of biological change based on human insight into DNA and RNA is also sustained by the first definition on biotechnology given in a recommendation of the Parliamentary Assembly of the Council of Europe in 1993: "Biotechnology which in a sense has a history as long as bread making and brewing can be defined as the use of biological organisms, systems and processes in industrial, manufacturing and service activities. The elucidation of the nature and functioning of the nucleic acids (DNA and

---

[36]The term is, of course, coined by Sir Thomas More in 1516, depicting "[a]n imaginary island (...) enjoying a perfect social, legal, and political system." (OED 1991, 370). His book is e.g published with a fine introduction as: More (1999). Further meanings listed in the OED 1991, 370-371 are: any imaginary, indefinitely-remote region, country, or locality; a place, state, or condition ideally perfect in respect of politics, laws, customs, and conditions; and, finally, an impossibly ideal scheme, esp. for social improvement, which is the sense in which I use the term here.

[37] Harwood and Wipat (2001), 65f.

RNA) in the 1950s has paved the way for the manipulation of the building blocks of living organisms so that cells or molecules can be altered."[38] Pivotal to this understanding of biotechnology are the notions of "production", "use", "manipulation", and "alteration", all of which presuppose an agent performing tasks intended to bring about the change within the biological (hyper-)structure of "organisms", "systems" and "processes" transcending the level of mere individual agency. At this point, it is important to stress that fundamentally speaking, the practices of biotechnology differ from those of technology in general by their trust in the ability of molecules to continue the processes instigated by the human agent.

The human interest in working with particular kinds of organisms or matter lies in their assumed response to the human act, which allows for carrying out such work in the hope of reaching goals that can only be achieved through this response. The more precisely this reaction can be anticipated, the more likely the outcome will match the expectation of the agent having caused it. While the human agency envisioned in this respect may be seen as an expression of human autonomy, the need for predictable response of the objects subjected to this agency may actually prove the assertive character of such a view. We shall return these questions in chapter 2.

In the Encyclopedia of Bioethics, the term is defined somewhat similarly, but more briefly than those presented so far: " Biotechnology' includes any technique that uses living organisms to make or modify products, to improve plants or animals, or to develop micro-organisms for specific uses."[39] Obviously, it is difficult to draw a very exact line between biotechnology and medicine, and if we take a statement from a representative of the industry, the connection becomes quite clear: "Modern Biotechnology and its application to human health care create an increasing number of innovative products and services, serving unmet medical needs."[40] In this sense, the very application of biotechnology can be understood as a medical deed, at least to the extent that it is used to satisfy the yet "unmet medical needs," for which traditional forms of medical practice have proven inadequate. This inadequacy is found whenever biotechnology promises to solve the problems it has identified, e.g. with regards to the improved effectiveness and profitability of antibiotics: "What in the 1950s to the 1980s

---

[38]Parliamentary Assembly of the Council of Europe: Recommendation 1213 (1993) on developments in biotechnology and the consequences for agriculture, point 1.

[39]Nowell (1995), 283.

[40]Tambuyzer (2000), 192-197.

was the working domain of the chemist, microbiologist and chemical engineer, has now in the 1990s encompassed the skills of the molecular biologist, enzymologist, protein chemist and the biochemical engineer. Biotechnology continues to contribute dramatic changes to the ways antibiotics are manufactured through fermentation, yield improvements, recovery processes and final product purity."[41]

Notwithstanding the actual design of biotechnology, especially with regards to its respective technical or practical side, as *technology,* it still is a field of applied sciences, and as such science, and therefore dependent upon observation and experiment as modes of relation describing the activity of an agent over against the objects of its research. This objectification presupposes a profound development, namely the separation of the human agent from its surroundings, in particular from nature, of which man himself nevertheless remains part. This separation is concomitant with the evolvement of human emancipation. Hence, I see biotechnology as the result of the following chain of movements:

human emancipation $\rightarrow$ science $\rightarrow$ technology $\rightarrow$ biotechnology

Consequently, I use the term 'biotechnology' in the broadest sense, subsuming issues of biomedicine and biotechnology alike: it is the name for practices in the fields of biology, chemistry and medicine employed with the purpose of modifying organisms capable of continuing a process instigated by a human agent using tools and insights especially produced by genetic and biochemical knowledge.

The often public concerns raised by biotechnology have instigated a subsequent discipline of counterbalance, namely bioethics, originally bred by theologians, nourish-ed by philosophers, and then growing into a distinct interdisciplinary field: "By the early 1970s, a genuine interdisciplinary conversation had blossomed, with its fruit a growing literature in which philosophy, theology, law, and other disciplines melded to describe, analyze, and advise on the questions of the new biology and the medicine."[42]

There is, of course, also a discussion of whether bioethics is an academic discipline. Albert R. Jonsen argues a little vaguely that "[i]n the strictest sense, it is not (...) A discipline is a coherent body of principles and methods appropriate to the analysis of some particular subject matter. Bioethics has no dominant methodology, no

---

[41]Lowe (2001), 350.
[42]Jonsen (1998), 84.

master theory. It has borrowed pieces from philosophy and theology." Understanding the notion of a discipline in this way, he admits, is "attractive, but probably an archaism [,]" because disciplines are "mosaics of theories,"[43] so that in the end, this discussion demonstrates that it is maturing as a discipline after all.[44] Others regard it as having arrived as an autonomous discipline already, deeming it "an unprecedented story of success."[45] The regularity of world congresses on bioethics is a further indication of this autonomy.[46] At the same time, bioethics is sometimes regarded as dialectically strengthening tendencies it originally was envisioned to counterbalance. In a nutshell, this criticism maintains that while the discipline contributes to socially close the conflicts arising with regards to biotechnology, it stimulates developments in this field, because it is "streamlining the different voices participating in public discussion (...)"[47] There is, in other words, at least allegedly a dominant discourse on ethics that "consists of a well-designed strategy of self-restriction."[48] The result is that new techniques are normalised. This criticism, which actually may be aimed at a specific kind of discourse, applying principlistic approaches in bioethics, has in turn provoked a discussion of its legitimacy.[49]

To my mind, this variety of interdisciplinary approaches and discussions in the field of bioethics is rather an argument for than against regarding it as discipline, since the whole idea of the academy as universitas indeed is collaboration on general and particular questions that have been arranged in fields of study and teaching, i.e. disciplines. This book reflects this view: I consider it to be a contribution to the field of bioethics, using methods of ethics and philosophy of religion to discuss some of the fundamental questions raised by biotechnology. It is not a presentation of the European

---

[43]Ibid., 345.

[44]Cf. ibid.

[45]ten Have (2001a), 1.

[46]Having met 2000 in London, the sixth world congress 2002 in Brasilia encompassed a number of national and international bioethics meetings, affirming the point made here.

[47]Kollek (2000), 156.

[48]Ibid.

[49]See e.g. Zwart (2000), 166: "...it is my contention that a legitimate and philosophically acceptable form of bioethics is possible, one that proceeds in a methodologically sound and well-considered manner and aims at recognising both the truths and the fallacies at work in scientific optimism as well as in public fear. The principlistic approach elaborated by Beauchamp and Childress and others often serves as a scapegoat for the kind of criticism articulated by Regine [scil. Regine Kollek]."

debate on bioethics, however, which can be found elsewhere[50], and neither a bioethical discussion of particular biotechnological practices. Instead, I see it as an attempt to contextualise the opaque concept of biotechnology in terms of its historical presuppositions, its systematic structure and current application in Europe.

2.2. Applications

Biotechnology is an expensive and exhaustive enterprise. In 2000, expenditures for research and development alone mounted to more than 4.9 billion € in the EU, with a venture capital investment of 3.8 billion €.[51] Especially small and medium sized companies, characterised by intensive research[52], are "structurally (...) very capital-intensive, and investments have long payback periods."[53]

During the 1980s, biotechnology was in Europe mainly carried out within large companies, but subsequently, small companies have been established, in 2001 amounting to a total of 1570 dedicated biotechnology companies with approx. 61,000 employees.[54] Still, "European biotechnology is (...) lagging significantly behind the US."[55] This is perceived as a deficit in European performance, which in part is related due to the late entry, with the most significant development taking place during the

---

[50]See e.g Dekkers (2001), 121.

[51]Bruno Hansen, Director Research DG European Commission: Biotechnology Research in Europe. The Sixth Framework Programme. Unpublished manuscript of speech delivered at the Belgian-Danish Forum for Innovation in Biotechnology. Hilton Brussels 29 May 2002, 6.

[52]See Commission of the European Communities (2001), 110: "...most European DBFs [scil. dedicated biotechnology firms, LR] are either micro or small research-intensive firms. Only approximately 10 per cent of active European DBFs have more than 50 employees, while the majority (about 57%) has less than 20 employees." As explained on the same page, the real number of small companies is presumably even higher, since the younger firms may be incorporated in alliances, venture capitalist etc.

[53]Commission of the European Communities: COM (2002) 27, 10.

[54]ibid., 9.

[55]Commission of the European Communities (2001), 127f. The number of companies is now higher in Europe than in the US, with 1570 registered firms compared to 1273 in the US. See Commission of the European Communities: COM (2002) 27, 9.

1990s[56], and in part due to a variety of structural factors.[57]

Since the mid 1970s, most corporate investments have been made in health care, in particular in pharmaceutical development.[58] It is in this field that the European Commission on a global scale identifies "a huge need (...) for novel and innovative approaches to meet the needs of ageing populations and poor countries."[59] This need exists, because cures for half of the world's diseases are still not known and existing cures increasingly lose their effectiveness due to evolving drug resistance, e.g. in terms of antibiotics. The Commission's anticipation that innovations in biotechnology will be able to meet these needs rests, clearly, on the results it regards as having been achieved already: "human growth hormones without the risk of Creutzfeldt-Jacobs Disease, treatment of haemophiliacs with unlimited sources of coagulation factors free from AIDS and hepatitis C virus, human insulin [as opposed to animal with severe side-effects, LR], and vaccines against hepatitis B and rabies."[60]

Most companies would focus on one field or a limited variety of diseases[61] by exploring possible causes at the molecular level, leading to e.g. the development of molecule drugs, the mapping of protein interactions, detecting of translocations, production of enzymes and hormone replacements.[62]

In terms of national economies, this field is less significant than that of food and

---

[56]Peak years for foundation of dedicated biotechnology firms (DBF) were 1997 an 1998. See for detailed graphs on distribution per country and year: Commission of the European Communities (2001), 108-112.

[57]Ibid., 127-130 presents a variety of difficulties, which can be summarised by quoting the very positive evaluation of the situation in the USA on p.128, according to which "US leadership in biotechnology derives from a unique blend of capabilities and institutional arrangements. These include a strong scientific, technological and industrial base; mechanisms that favour communication and transfer of knowledge between academia and industry; a financial system that promotes the start-up of new, risky ventures; strong intellectual property protection [scil. patents, LR]; and a favourable climate in terms of public perception and regulation that does not restrict genetic experimentation." It is pointed out, however, that the US model does not have to be followed necessarily.

[58]Bains and Evans (2001), 255.

[59]Commission of the European Communities: COM (2002) 27, 5.

[60]Cf. ibid.

[61]Such as infectious diseases, allergies, multiple sclerosis, leukemia, diabetes, haemostasis, arthritis, neurodegenerative diseases and cancer.

[62]These are just some of the areas presented by Danish biotechnology companies in: The Danish Trade Council. Royal Danish Ministry of Foreign Affairs. Belgian-Danish Forum for Innovation in Biotechnology. Hilton Brussels Boulevard de Waterloo. 29 May 2002. S.locus. 9-16.

agriculture, which nevertheless attracts fewer companies, because bluntly put, "a new food cannot be sold at $1000 a meal in the same way that a new drug can be sold at $1,000 a bottle."[63] There are exceptions within this sector, though. Recoupment of costs can be obtained, e.g., in breeding of crops and animals, which has instigated the development of new strains of plants as well as new animal reproduction techniques, such as cloning and transgenetic engineering.[64]

Generally speaking, the genetic modification (GM) of plants is envisioned as a means to solve the most likely increasing problem of providing food for a rapidly growing world population, the majority of which will live in poverty. GM plants are designed to resist natural calamities better, reducing the use of chemical pesticides[65], fertilisers and drugs, while increasing conservation tillage, leading to "more sustainable agricultural practices."[66] The mapping of the rice genome may in this regard prove to be of very high significance in order to allow for developing robust forms of this plant.[67] Therefore, a leading expert in the field, Marc van Montagu, exclaims: "Yes, we need GM plants and the tools of Biotechnology urgently."[68]

Genetically engineered enzymes are used e.g. for food processing and garment treatment. Other examples of application are the use of micro-organisms in the paper pulp and plastics industry.[69] Thus, one may encounter results of biotechnological techniques in the ordinary and extraordinary situations of one's life, sometimes even without being aware of it.

In all of these sectors, the actual techniques basically deal with the molecules of cells, such as lipids, nucleic acids, proteins, and phosphates. Since cells are viable, developing through the interaction of genetic information with an environment, and

---

[63] Bains and Evans (2001), 259.

[64] Ibid.

[65] "Pesticides in general pose significant health risks for people exposed to them, especially children, and even unborn infants. Pesticides have been shown to affect reproductive cells and processes in other animals; if reproductive processes are affected in humans to the same extent, then the pesticides used today have the potential to impact future generations of human beings decades from now." Pimentel and Hart (2001), 97.

[66] Commission of the European Communities: COM (2002) 27, 6.

[67] Cf. Science, April 2002

[68] In his presentation at the first Belgian-Danish Forum for Innovation in Biotechnology, Brussel 29 May 2002.

[69] Bains and Evans (2001), 259f.

able to reproduce, cells are the most elementary forms of micro-organisms. With regards to micro-organisms, C. Ratledge formulates what could be seen as a first law of biology, namely that *the purpose of a micro-organism is to make another micro-organism*. Hence, the biotechnologist exploits the micro-organism in order to either produce as many of them as possible or to use them in order to produce something which he desires by way of diverting the reproductive capacity towards that goal.[70] In accordance with our definition of biotechnology given above, this principle could also be applied to biotechnology in general: a technique is used by a human agent in order to modify an organism, i.e. achieving a goal by making use of that organism's ability to continue the process instigated, which is why humans e.g. take medicine or sow.

In spite of its many fields of application, biotechnology is in the media mostly perceived as relating to the genetic engineering of food, animals, and humans, and less with the development of e.g. medication apart from its use of animals for trials. Particular concerns have been raised with regards to actual or potential[71] techniques subsumed under the headings of cloning, genetic modification of products and genetic testing.

As I here venture to illuminate the situation of biotechnology as if surpassing, i.e. seeing it the way it simply is, I also need to account for its actual application. I shall now do so by focussing on one of these debated topics, namely cloning, giving special attention to the new field of stem cell research, which, as you will see shortly, also is related to two of these debated fields. The idea is to use this as a *pars pro toto*, explaining the techniques used and identifying some of the issues raised in conjunction with them. This limitation to one topic is also necessary, because a presentation of all current biotechnological techniques would make little sense in view of the detailed reference works now available. Nevertheless, other procedures will feature, albeit briefly, whenever required for contextualisation in the following chapters.

The word "cloning" is commonly used for the copying of cells. This technique is used in different settings. For instance, in genetic engineering, the object of engineering

---

[70]Cf. Ratledge (2001), 17.

[71]The notion of potentiality is in this respect a little vague. As we do not *know* the future, we cannot fully conceive the way in which anything will be used either. Furthermore, the more complex a matter is, the higher is the risk of misinterpreting the facts we encounter. Therefore, things and beings may hold potentials of which we are quite unaware, which means that fears surrounding the potentiality of present techniques may seem unwarranted in the light of actual possibilities, but not necessarily so as far as the potential for future use is concerned. We shall briefly return to this question in chapter 2, 2.

is the basic component of DNA, nucleic acid. By using bacteria as hosts, identical copies of fragments of DNA can be produced, i.e. cloned, which forms one important cornerstone of this field.[72] The techniques of such molecular cloning have "led to the production of such important medicines as insulin, growth hormones, erythropoietin (necessary to treat anemia associated with dialysis for kidney disease) or tissue plasmogen activator (tPa) to dissolve clots after a heart attack."[73]

Cloning is also used in plant breeding and it occurs spontaneously e.g. in the case of vertebrates and mammals, normally called 'twinning'. These forms of cloning do usually not stir public emotion, but reproductive cloning by somatic cell nuclear transfer (SCNT) does. This technique "consists of replacing the ovum's haploïd nucleus by a diploïd coming from a differentiated somatic cell, originating from a child or an adult individual."[74] In other words, the nucleus of an unfertilised eggcell, i.e. its DNA, is replaced with the DNA from the nucleus of a cell that is already developed, i.e. functioning in its genetically preprogrammed place. This somatic cell is diploïd, of course, because it has developed after fertilisation. Thus, there is only parent in the case of SCNT. The sheep called 'Dolly' resulted from the use of this technique after 277 attempts.[75] In the aftermath of its presentation, the use of this technique for artificially creating human twins has been banned internationally.[76] With the cloning of a monkey in 2000, some see research nevertheless moving towards human cloning, further fuelled

---

[72]Harwood and Wipat (2001), 66f.

[73]Bart Hansen/Paul Schotsmans: Stem Cell Research: A Theological Interpretation. Manuscript of forthcoming publication, 2.

[74]Ibid., 3.

[75]A brief description of this process in biological terms is found in: Bart Hansen/Paul Schotsmans: Stem Cell Research: A Theological Interpretation. Manuscript of forthcoming publication, 3.

[76]The Universal Declaration on the Human Genome and Human Rights, adopted by the General Conference of UNESCO at its 29th session (1997), declares in art. 11: "Practices which are contrary to human dignity, such as reproductive cloning of human beings, shall not be permitted." The additional protocol to the Council of Europe Convention on Human Rights and Biomedicine on the Prohibition of Cloning Human Beings, opened for signature on 12 January 1998, states the prohibition in very clear terms in article 1: "Any intervention seeking to create a human being genetically identical to another human being, whether living or dead, is prohibited." The protocol has been signed by 29 countries, of which 13 have ratified it. The Charter of 28 September 2000 on Fundamental Rights of the European Union, approved by the European Council 14 October 2000, prohibits in article 3 "the reproductive cloning of human beings".

by researchers who provoke by claiming to break the ban.[77]

SCNT is also used for so-called therapeutic cloning techniques, where the aim is to treat specific diseases related to tissue degeneration rather than develop an embryo in vitro. The technique of cloning can therefore be used in order to achieve varying goals. Accordingly, it has been proposed to speak of (human) reproductive cloning and nuclear transplantation respectively, in order to mark the difference in objectives.[78]

In terms of using nuclear transplantation for treating human diseases, high hopes have been placed in the so-called (human) stem cells, because of their ability to "divide to produce either cells like themselves (immortality), or cells of one or several specific differentiated types (potentiality)."[79] For the European Commission, they offer the prospect of producing tissues and organs in order to treat degenerative diseases and injuries related to strokes, Alzheimer's and Parkinson's diseases, burns and spinal-cord injuries.[80] Research on these cells intensified after the US government in August 2001 decided to federally fund research on 64 stem cell lines derived from embryos, which allowed university laboratories to engage in this work.[81]

Commonly, yet not unanimously, stem cells are defined as "cells with the capacity for unlimited or prolonged self-renewal that can produce at least one type of highly differentiated descendant."[82] These cells exist throughout the stages of human development, but only embryonic stem cells (ES cells) possess pluripotentiality, i.e. the potential to develop into any cell type of a human adult, and so-called immortality, i.e. here to remain undifferentiated for a longer period. This latter feature allows for various forms of genetic engineering, in which e.g. a particular gene could be changed or added while the cell is kept in a petri-dish. ES cells are not totipotent, however, i.e.

---

[77]The news of this result were published 14th January 2000. The monkey was cloned, i.e. split from an embryo after 107 embryos had been divided into two or four, resulting in 368 embryos in total, of which only one developed successfully.

[78]See e.g. Bart Hansen/Paul Schotsmans: Stem Cell Research: A Theological Interpretation. Manuscript of forthcoming publication, 4, with reference to P. Verspieren: Le clonage human et ses avatars, in: Etudes 391 (1999), 459-467 and B. Vogelstein/B. Alberts/K. Shine: Please don't Call It Cloning, in: Science 295 (2002), 1237-1238.

[79]Bart Hansen/Paul Schotsmans: Stem Cell Research: A Theological Interpretation. Manuscript of forthcoming publication, 4.

[80]Commission of the European Communities: COM (2002) 27, 6.

[81]Cf. Bart Hansen/Paul Schotsmans: Stem Cell Research: A Theological Interpretation. Manuscript of forthcoming publication, 7.

[82]Watt and Hogan (2000),1427.

they cannot develop into an embryo on their own. The term 'embryonic' is used, because these cells are typically derived at the blastocyst stage, i.e. the developmental stage five to six days after fertilisation of the ovum, where a cell cluster has been arranging from which placenta, foetus and other tissue will be formed.[83] Since the ES cells will become more specialised in the subsequent development, they will gradually lose their pluripotentiality.

In 2000, F. Watt and B. Hogan pointed out that the "spotlight on stem cells has revealed gaps in our knowledge that must be filled if we are to take advantage of their full potential (...)"[84] Since then, only some of these gaps have been filled and there still is a disagreement on whether to focus research on multipotent adult or pluripotent embryonic stem cells, depending on the ethical concerns identified with either. Therefore, research on adult stem cells has been intensified in the quest for embryo-saving alternatives due to the prevalent "technical obstacles of the human cloning combined with the ethical controversy about using embryos for research purposes[.]"[85] At the level of the European Union, the European group on ethics to the commission "stresses in particular that there should be European Union funding for research into adult stem cells. Such research is less attractive than other types of research for private investors given the difficulty of isolating these cells, but it does not raise the same ethical objections as does the removal of stem cells from human embryos."[86] The ethical objections referred to are raised, because the use of ES cells is inherently linked with the question of obtaining them. For in order to use embryonic stem cells, an embryo is needed from which these cells can be removed. Such embryos now typically stem from assisted fertilisation procedures, where among the number of embryos resulting from in-vitro-fertilisation, some or even most may not be implanted after all. They are often referred to as "*spare* [my italics, LR] embryos from in vitro fertilization procedures,"[87] which is not an altogether neutral description. The reason for using this term is that for IVF, several ova will be fertilised, since nidation often proves quite

---

[83]Cf. Bart Hansen/Paul Schotsmans: Stem Cell Research: A Theological Interpretation. Manuscript of forthcoming publication, 6.

[84]Watt and Hogan (2000), 1427.

[85]Cf. Bart Hansen/Paul Schotsmans: Stem Cell Research: A Theological Interpretation. Manuscript of forthcoming publication, 5.

[86]European Commission (2001), 6.

[87]Watt and Hogan (2000), 1427.

cumbersome to achieve. It can be anticipated, therefore, that not all embryos will be needed, but not precisely determined how many will be left. The term 'spare' thus means 'not needed for artificial fertilisation', but could, misleadingly, also be understood to refer to the quality of the human embryo.[88] Given that research in this field is new, is not surprising that at the national European level, "stem cell research is not regulated as such."[89] Some countries have begun legislative procedures[90], while the Council of Europe convention on biomedicine and human rights prohibits it implicitly in its art. 18. At least the issue of ES cells clearly demonstrate how research in biotechnology advances faster than the concomitant societal reflection on its use and legal status.

---

[88]The European Group on Ethics defines it in this way: "spare embryos' (i.e. supernumerary embryos) created for infertility treatment to enhance the success rate of IVF, but no longer needed for this purpose. They are intended to be discarded, but instead, may be donated for research by the couples concerned[.]" Opinion of the European Group on Ethics No 15 (2000), 127.

[89]Opinion of the European Group of Ethics No 15 (2000), 128.

[90]France and Germany have legalised the import, but not the creation of ES cells.

CHAPTER 2

COUNCIL OF EUROPE POLICIES ON BIOTECHNOLOGY

If biotechnology indeed is an entity of diverse means and objectives formed in the nexus of academic, political, and economic interests, it is necessary to locate this nexus. Such an endeavour surpasses merely demonstrating that e.g. research in biotechnology is being carried out at universities or that particular practices of biotechnology are employed in order to generate products intended for sale, for a quick internet search, a phone call or a site visit could prove this in instants. Instead, the idea is to give an impression of the values, goals and institutions shaping the European debate on biotechnology in order to let the distinct contours of biotechnology emerge, for public deliberations and policies contribute quite as much to the actual state of biotechnology as the techniques employed in its name.

I limit this survey to the documents issued by the Council of Europe (COE), because it is the most extensive European workshop of transnational policies, reflecting various levels of multilateral agreement reached on how to shape biotechnology in Europe. Choosing the COE may seem ideosyncratic, but is to my mind at least warranted from a formal point of view. If we do take the idea of democracy seriously, we have to assume that the national governments and parliaments represent the will of the electorate. Since the COE is the only pan-European organisation that unites those representatives, albeit through additional acts of selective representation by delegation, we can assume that the opinions expressed in this setting formally speaking express national views formed by some societal agreement in the actual member state. Since any political deliberation is subject to the general mechanisms of human negotiation, there is no guarantee for the full representativeness of the positions taken.

The European Union is, of course, also a major institution in this regard, with e.g.

economic policies issued on our topic[91], but it involves only fifteen states[92] at present and the COE has a longer tradition of reflections on the underlying principles and values of the European legal system. In its own confident words, the COE "is the only organisation which has a pan-European dimension and well-established competence in the fields of human rights and ethics."[93]

There are substantial differences in the contents, scope, and impact of their policies due to their respective institutional backgrounds: the COE is an organisation of sovereign states, deliberating and deciding through their representatives on matters they deem fit for common action, typically concerned with issues of more fundamental character, such as the protection of human rights.

The European Union (EU) once started off with six states interested in coordinating their production of coal and steel (ECSC),[94] which soon was extended to their use of atomic energy (Euratom)[95] and to cooperation within a wider economic community focussed on easing trade by ensuring the free movement of goods and

---

[91]As the latest example, see e.g. the rationale for the 6th framework programme: "The Commission aims to restore European leadership in life sciences and biotechnology research. The 6th Community Framework Programme for Research, Technological development and Demonstration activities (2002-2006) proposes this area as the first priority and will provide a solid platform for constructing, in collaboration with the Member States, a European Research Area."Commission of the European Communities: COM (2002) 27, 8. The financial resources reserved for this project are considerable, e.g. 2,2 billion € for genomics and biotechnology for health and 685 million € for food quality and safety. 15% of the budget is reserved for the SMEs, i.e. small and medium sized businesses. See Bruno Hansen, Director Research DG European Commission: Biotechnology Research in Europe. The Sixth Framework Programme. Unpublished manuscript of speech delivered at the Belgian-Danish Forum for Innovation in Biotechnology. Hilton Brussels 29 May 2002, 15.

[92]These states have acceded 1st January of the following years: Austria (1995), Belgium (1957), Denmark (1973), Finland (1995), France (1957), Germany (1957), Greece (1981), Ireland (1973), Italy (1957), Luxembourg (1957), Netherlands (1957), Portugal (1986), Spain (1986), Sweden (1995) and the United Kingdom (1957).

[93]Council of Europe, Committee of Ministers: Biotechnology and intellectual property. Reply from the Committee of Ministers to the Recommendation 1425 (1999), adopted at the 730th meeting of the Ministers' Deputies (22 November 2000), Doc. 8894. Strasbourg 27 November 2000, point 3.

[94]The Treaty on the European Coal and Steel Community (ECSC) was signed by Belgium, France, Germany, Italy, Luxembourg, Netherlands in Paris in 1951.

[95]The Treaty on the European Atomic Energy Community (EAEC or Euratom) was signed by the same states in Rome in 1957.

workforce, the EEC.[96] In 1967, these organisations merged into the 'European Communities' (EC). While the idea of an actual union of states was present *in nuce*, and national sovereignty has been gently dissembled through the course of years, this political ambition first became fully translucent during the 1990s. Several charters and treaties have paved the way to the manifestations of a union that formally entered into force 1 November 1993 (henceforth: EU) and of which the subsequent abolishment of internal border control[97] and the introduction of a single currency[98] are only the beginning.

A certain rivalry between the two institutions may be observed, especially whenever the COE primarily is pictured as an instrument of traditional European diplomacy or the EU primarily as an instrument of economic administration.[99] It is also not uncommon to experience that one institution is taken for the other, facilitated by the variety of institutions under different auspices sometimes even residing in the same city as in the case of Strasbourg.[100]

---

[96]The Treaty on the European Economic Community was likewise signed by these states in Rome in 1957. Cf. for a brief and very clear presentation of the EU structures: M.A. Theofilatou: The emerging health agenda. The health policy of the European community. Doctoral dissertation at the University of Maastricht. s.l. 2000, especially pp. 24-74.

[97]The Schengen Agreement came into force between Belgium, France, Germany, Luxembourg, the Netherlands, Portugal and Spain 26 March 1995. It abolished formal border control, replacing it by random checks at and near the borders. Austria signed the agreement the same year,

[98]The Euro, formally introduced as a currency 1 January 1999 and as money 1 January 2002 in Austria, Belgium, Finland, France, Germany, Italy, Ireland, Luxembourg, Netherlands, Portugal, Spain. By special agreement, the Euro is also introduced in states having had moentary unions with current Euro countries, namely Monaco, Andorra, San Marino, and the Vatican State, of which the latter may mint.

[99]Sometimes, this can also be sensed in the field of ethical reflection. When describing its own set-up, the European Group of Ethics, appointed by the European Commission, stresses its independence as opposed to e.g. the "Council of Europe's Steering Committee on Bioethics (CDBI) (which drafted the convention on Human Rights and Biomedicine) (...) composed of members who are delegated by their governments and therefore have the task of presenting their positions." European Group on Ethics in Science and New Technologies to the European Commission: General Report on the activities of the European Group on Ethics in Science and New Technologies to the European Commission 1998-2000. Luxembourg: Office for Official Communications of the European Communities 2001, 3.

[100]The Council of Europe with its parliamentary assembly and committee of ministers resides here, as does approx. four days a month the European Parliament, which otherwise gathers in Brussels. The European Court of Human Rights is in Strasbourg (not to be confused with the

The COE is an intergovernmental organisation. After signature of the statutes in London on 5 May 1949, it was inaugurated on 8 August the same year, when the Committee of Ministers representing Belgium, Denmark, France, Ireland, Italy, Luxembourg, the Netherlands, Norway, Sweden, and the United Kingdom first met in Strasbourg. At this meeting, it was decided to invite Greece, Turkey and Iceland, of which the first two were welcomed the next day and the last few months later. Thirty states joined after a variety of domestic and accessory processes, with forty-one states currently represented on the council[101]: Cyprus (1961) and Malta (1965) obtained their independence; The Federal Republic of Germany (1950 associate member as the Saarland, 1956 full member) and Austria (1956) regained sovereignty, albeit with allied stipulations; Portugal (1976) and Spain (1977) overcame their regimes of isolation and oppression. Switzerland (1963) took first steps towards international collaboration. After the fall of wall first the already officially independent states of Finland (1989), Hungary (1990), Poland (1991) Bulgaria (1992), Czech Republic and Slovakia (1993, after having joined between 1991 and 1992), Romania (1993), and Albania (1995). This group was complemented by the states emerging from the broken Sovjet Union: Estonia (1993), Lithuania (1993), Latvia (1995), Moldova (1995), the Ukraine (1995), Russia (1996) and Georgia (1999). In the course of the painful disintegration of former Yugoslavia, Slovenia (1993), the 'former Yugoslav Republic of Macedonia' (1995), and Croatia (1996) as the latest so far, with Bosnia and Herzegovina having applied for membership. In addition, the smaller European states of Liechtenstein (1978), San Marino (1988) and Andorra (1994) joined after having installed the in this regard necessary bodies for cooperation.

The aim of the Council of Europe (COE) "is to achieve a greater unity between its members, in particular through harmonising legislation on matters of common interest."[102] This is done through the three main bodies of the COE, the parliamentary

---

International Court of Justice in the Hague), while the Court of Justice of the European Communities resides in Luxembourg. The European Commission and the European Council of Ministers have their headquarters in Brussels.

[101] An investigation of the accessory process of each country is less relevant in our present context, but it would form an interesting task. Details of the processes can be found in the published debates of the Parliamentary Assembly during the various periods.

[102] E.g. Council of Europe, Committee of Ministers: Resolution (78) 29 on harmonisation of legislations of member states relating to removal, grafting and transplantation of human substances, preamble.

assembly, the council of ministers and the European Court of Human Rights.[103] The assembly is formed of members elected by the national parliaments, "representing public opinion through European Parliaments."[104] It offers recommendations for or opinions on measures to be taken by the committee of ministers, while it only may invite member states to act. In other words, it is the deliberative body of the COE. The committee, "representing Governments"[105], decides by way of resolutions or recommendations addressed to the member states. Neither the assembly nor the committee has, thus, any legislative power, since that rests solely with the national states.

The main legal instruments of the COE are conventions, which in fact are international treaties signed and ratified by sovereign states. A convention may be ratified, accepted or merely approved by a member state, with or without reservations made with regards to certain provisions. It reflects the modern idea of the *contrat social* in that a body, here the state, freely can bind itself to a law uniting all bodies on the grounds of common insight.

The spirit and letter of the conventions are protected by the European Court of Human Rights in the following ways: a case alleging the violation of the convention of human rights can be heard, if this case is declared admissible. Decisions on matters relating to subjects in that convention may be informed by other generally accepted European treaties. The court can also give advisory opinions on the interpretation of a convention, if this is stipulated in it, e.g. as in the convention on human rights and biomedicine. Furthermore, disputes concerning COE conventions can also be brought before the International Court of Justice in The Hague, if all parties involved have recognised the jurisdiction of that court.[106]

---

[103]The now single court of human rights replaced in 1999 the two-tier structure of the (European human rights) commission and the European court of human rights. This reorganisation was intended to improve administration: "Doch diese Strategie ist offensichtlich fehlgeschlagen und für die bislang zugelassene Zahl von 1000 Fällen für das erste Halbjahr 1999, die nur einen geringen Prozentsatz der insgesamt eingereichten Menge beinhaltet, ist bereits mir (scil. 'mit', LR) einer Wartezeit von fünf Jahren zu rechnen bis es zu einer Verhandlung kommt." Schwinger (2001), 81.

[104]Council of Europe, Directorate of Information: European Convention on Human Rights. Strasbourg 1953, 3.

[105]Ibid.

[106]I am indebted to Dr. Peteris Zilgalvis, COE DG I, Bioethics Division for clarification on this matter.

Responding to the challenge of the United Nations Declaration of Human Rights of 10 December 1949, the European Convention for the Protection of Human Rights and Fundamental Freedoms (ECHR) was signed in Rome 4th November 1950.[107] It may be seen as the founding charta of the COE, providing the necessary legal grounds for defending pluralist democracy, respect for human rights and the rule of law as anteceding principles, perhaps rooted in natural law. The Heads of State and Government affirmed these principles in 1997 and underlined the "standard setting role of the Council of Europe in the field of human rights and its contribution to the development of international law through European Conventions (...)"[108] In a brochure on the ECHR published in 1953, the COE links its aim, "the achievement of greater unity between its members", with historical experience when pointing out that this unity is "also furthered by protecting Human Rights by making them stronger where they are found to be weak so that they may withstand the peril of dictatorship should it ever arise."[109] In short, liberty is the foundation of peace and justice, requiring democracy and observance of human rights, which the COE safeguards by creating unity e.g in legal matters between the European states in order to bar dictatorship.

Within the COE framework, the discussion of biotechnology is situated in the broader context of its positions on bioethical matters. The assembly initiated this reflection in 1976, wishing to solve the tension between "a zealous application of the most modern techniques for prolonging life" and "the true interests of the sick (...)"[110] What started as a reflection on a fairly distinct question of medical ethics in the 1970s

---

[107]Council of Europe, Directorate of Information: European Convention on Human Rights. Strasbourg 1953, 3. The convention itself is published as: Council of Europe: Convention for the Protection of Human Rights and Fundamental Freedoms as amended by Protocol No. 11, Rome, 4.XI.1950. European Treaties Series (ETS) 5.

[108]Second Summit of the Council of Europe: Final Declaration. Strasbourg 11 October 1997, in: Council of Europe, Directorate of Legal Affairs: Texts of the Council of Europe on bioethical matters. CDBI/INF (98) 5. Strasbourg 1998, 101-108, 101.

[109]Council of Europe, Directorate of Information: European Convention on Human Rights. Strasbourg 1953, 3. The brochure is popular in tone, presenting the main structures and aims of the COE very clearly. It is also outspoken: "...Hitler and other dictators, past and present, showed that human rights and fundamental freedoms can be jettisoned in less than no time. Today action is needed if human rights are to be secured." Ibid., 2.

[110]Parliamentary Assembly of the Council of Europe: Resolution 613 (1976) on the rights of the sick and dying, point 1.

soon became an attempt to grasp the possibilities, advantages and dangers first of genetic engineering in the 1980s, then of the whole field of biomedical sciences and biotechnology, realising its universal problems and need for solutions to be found with the aid of the new discipline of bioethics in the 1990s, resulting in the 1997 convention for the protection of human rights and dignity of the human being with regard to the application of biology and medicine and its subsequent protocols on specific problems from 1998 onwards. Since 1999, where the COE also returned to the matter of the terminally ill and dying, increasingly specific recommendations on biotechnology and xenotransplantation are being addressed also to wider audiences, including the European communities and global governmental and non-governmental organisations. This development is concomitant with the above mentioned way in which biotechnology is emerging as an international field of converging interests, research and initiatives.

Generally speaking, the committee of ministers have approached the field of biotechnology, which according to our definition covers the fields of biomedicine and biotechnology alike, by recommending harmonisation of legislation on specific biotechnological procedures with regards to the ethical, legal, and social implications given by their application. Since its resolution on the "harmonisation of legislations of member states relation to removal, grafting and transplantation of human substances"[111] in 1978 and its sequel in 1979, recommendations have mainly been made in the 1990s for the new techniques involving human tissue, blood, DNA, xenotransplantation, and, of course, for the preparation of the convention on biomedicine, in addition to an elaborate set of rules for medico-legal autopsy. Most recently, it has been dealing with waiting lists and waiting time in conjunction with transplantation in 2001, returning to the topic of the late 1970s.

1. THE PARLIAMENTARY ASSEMBLY

Dealing with matters in biotechnology, the assembly acts on the presupposition that "the exploitation of technological opportunities not only in science but also in medicine

---

[111]Council of Europe, Committee of Ministers: Resolution (78) 29 on harmonisation of legislations of member states relating to removal, grafting and transplantation of human substances. Adopted by the Committee of Ministers

must be governed by clear ethical and social guidelines (...)"[112] The ambiguity thus referred to, albeit indirectly, is profoundly marking human existence: *any* use may be abused, turning opportunities into dangers, the evidence of which is expressed in the traditional principle of *abusus non tollit usum*.

When encountering opportunities not anticipated, the most likely human reaction is either rejection or fascination, approaches often a little pejoratively characterised as 'traditional' or 'progressive'. This pattern of reaction also emerges on the COE level.

In our context, it is interesting to examine how the COE assembly of national parliamentarians has sought to 'govern' the 'opportunities', revealing contemporary views on biotechnology as well as the underlying notions forming them.

1.1. The beginnings

In its first proper document on a theme of biotechnology in 1982, the assembly was enthused by genetic engineering:

> "i. the techniques of genetic engineering present an immense industrial and agricultural potential which in coming decades could help to solve world problems of food production, energy and raw materials; ii. radical breakthroughs in scientific and medical understanding (university of the genetic code) are associated with (...) these techniques; (...) genetic engineering holds great promise for the treatment and eradication of certain diseases which are genetically transmitted."[113]

The hyberbolic terms "immense potential", "solve world problems", "radical breakthroughs", "great promise" and "eradication" indicate a fundamental trust in the human possibility to overcome contemporary deficiencies in nature, irrespective of its sentient status. The wording "eradication of certain diseases" reflects an anteceding act of defining genetic disease almost in ontological terms, i.e. as existing independently of the person affected, after which the goal of eradicating this privation or deformation can be set, for otherwise, one would have to equal "eradication of disease" with "eradication of the person carrying it". Disease does not exist as such, however, i.e. not as *the* 'disease', but always as the actual state of a person in relation to a former or an

---

[112]Parliamentary Assembly of the Council of Europe: Recommendation 1046 (1986) on the use of human embryos and foetuses for diagnostic, therapeutic, scientific, industrial and commercial purposes, point 4.A.

[113]Parliamentary Assembly: Recommendation 934 (1982), 15f.

anticipated one, i.e. the 'disease of that person' or, more precisely and less pejoratively put, the 'state of that person.' Apart from objectification, the term 'disease' is also a means of classification and even separation, whenever some define a group of people as in need of treatment, distancing themselves from them by marking them as 'other'. We shall return to this field when discussing the modern views on the modern concepts of agency and identity in chapter 2, section 3.

In spite of its enthusiasm for the new techniques, the assembly recognises risks attached to research in this field. Defending the right to inherit a genetic pattern without change,[114] it presents the first idea of a European convention on the new technologies by suggesting this right "be made explicit in the context of the European Convention on Human Rights" and the committee of ministers "draw up a European agreement on what constitutes legitimate application to human beings (...) of the techniques of genetic engineering."[115]

1.2. Developments of doubt and difficulty

Four years later, the tones of fascination are mixed with those of rejection in its recommendation 1046. The "recent progress in the life sciences and medicine, in particular in animal and human embryology" is still qualified as "remarkable" and in vitro fertilisation (IVF) as an achievement, for by it, "man has (...) the means of intervening in and controlling human life in its earliest stages [.]" In spite of this positive view on the human maker, however, clear ethical and social guidelines are needed and "future benefits from the advance of medical science and technology must be carefully assessed in deciding when, and how, and on what grounds, to restrict the exploitation of technological opportunities."[116]

The reason for this slight change in attitude appears later in the document, where the assembly recommends that the committee call on the governments of the member states to "investigate the rumours about a trade in dead embryos and foetuses circulating in the media, and to publish the results (...)"[117] Furthermore, the use of human embryos and foetuses and materials stemming from them should be limited to

---

[114]Ibid., 17.
[115]Ibid.
[116]Parliamentary Assembly: Recommendation 1046 (1986), 19.
[117]Ibid., 20.

"strictly therapeutic" purposes, and any deviation forbidden, which include cloning, selection of sex, creation of chimeras, ectogenesis, research or experiments on living human embryos and "the implantation of a human embryo in the uterus of *another* (my italics, LR) animal or the reverse."[118] This last stipulation bears, perhaps unwillingly, testimony to a profound difficulty encountered by the assembly: it could not define the human being, which would have been required, since scientific progress had made it possible to "intervene in developing human life from the moment of fertilisation (...)" Instead, differing views are offered in the document: "from the moment of fertilisation of the ovule, human life develops in a continuous pattern," but "it is not possible to make a clear-cut distinction during the first phases (embryonic) of its development, and (...) a definition of the biological status of an embryo is therefore necessary."

Considering the legal status of the embryo and foetus as not having been defined by law yet, the assembly comprehends "the variety of ethical opinions on the question of using the embryo or the foetus or their tissues, and conflicts between values which arise [.]" At the same time, "human embryos and foetuses must be treated in all circumstances with the respect due to human dignity (...)"[119] There is little logic in such a compromise, for if human life indeed develops in a *continuous* pattern and if embryos and foetuses must be treated with the respect due to human dignity, i.e. the dignity humans inherently possess, it is clear that these entities are not merely humanlike, but indeed human, which in turn should clarify their legal status.

Again, the idea of a Convention "or any other suitable legal instrument" is mentioned as a means of responding to the problems in "relation to the use of human embryonic and foetal tissue"[120]

This question of definition reappears in 1989, where the assembly struggles to establish "the necessary balance between the principles of freedom of research and of respect for human life and other aspects of human rights"[121] Taking a clearer stance than in recommendation 1046 from 1986, the assembly considers it "appropriate to determine the legal protection to be given to the human embryo from the time that the human egg is fertilised, as foreseen in Recommendation 1046." As just pointed out, the recommendation does not precisely foresee this. Notice that the legal protection is "to

---

[118]Ibid., 21, with the extensive list of the procedures desired banned.

[119]Ibid., 20.

[120]Ibid., 22.

[121]Parliamentary Assembly: Recommendation 1100 (1989), 25.

be given", i.e. extrinsically determined, rather than recognised as demanded by its status. Moreover, a minor confusion remains, "[c]onsidering that the human embryo, though displaying successive phases in its development (...), displays (...) a progressive differentiation as an organism and none the less maintains a continuous biological and genetic identity."[122]

So, the identity and protection of the human embryo is contrasted with its differentiation as an organism, paying tribute to the opposed interests of defending the human status of the embryo as well as the progress of science and technology. This conflict is already present at the very beginning of the document, where the word 'biotechnology' appears for the first time in an assembly recommendation: "Considering that science and technology, and especially the biomedical sciences and biotechnology, continue to advance and develop as an expression of human creativity, and that their freedom of action cannot be restricted arbitrarily, but only on the basis of, *inter alia*, professional, legal, ethical, cultural and social principles for the protection of human rights and the dignity of man as an individual and social being;"[123]

Biotechnology is, thus, related to technology in the same way that the biomedical sciences are related to science, with technology here understood as an application of science, much in the sense of the standard definitions discussed earlier. Furthermore, the notion of 'creativity' is used to qualify the whole field as positive, while the notion of 'freedom' indicates a fundamental right, that "cannot be restricted arbitrarily[.]" Such freedom of action requires an agent able of acting freely and who in this also is entitled to do so. In other words, the mere concept of human freedom warrants its own right of execution: since free, humans are entitled to *act* freely. In turn, one would have to ask at what point restricting human acts may affect the inherent freedom of the individual in such a way that it ceases to exist, to which we shall return in chapter 3.

Still, a view of freedom as *completely* unrestricted would be difficult to defend, last but not least in terms of the rule of law, which according to article 3 of the COE Statutes is the complementary principle to the human rights and fundamental freedoms enjoyed by all persons.[124] The range of principles listed here is envisioned as means of

---

[122]Ibid.

[123]Ibid., 24.

[124]COE Statutes article 3: "Every member of the Council of Europe must accept the principles of the rule of law and of the enjoyment by all persons within its jurisdiction of human rights and fundamental freedoms."

protecting human rights and dignity. Ironically, freedom could be restricted, then, in the interest of protecting the human right of, say, freedom. The important qualifier used for such restriction is the term 'arbitrarily', i.e. acting without due cause. This term describes behaviour considered unethical due to the inequality it causes: the intended unpredictability of the arbitrary act degrades the person subjected to it, since the motivation and goal of the act is veiled by the person pursuing superiority. Using the term in this context, the assembly denounces any *unethical* restriction of freedom, while acknowledging the legitimacy of restricting it on the basis of principles considered *ethical*. This hardly surprising position is also inspired by the wish to defend the quite diverse interests of progress, harmony, liberty and social justice, the balance of which is sought by the custom of constantly adapting legislation to new knowledge and existing values.[125]

The assembly remains torn between fascination of the new possibilities and the rejection of the risks attached to them: the comprehensive list of research and experimentation permitted on human gamets, embryos and foetuses in the appendix to the document, which almost grants total freedom to research using genetic or recombinant genetic material,[126] is anteceded by a call to "establish, as a matter of urgency, as a safeguard, an international multi-disciplinary body to ensure convergent approaches (...) and to avoid (...) the creation of "genetic havens.""[127] Bearing in mind the very cautious style characterising COE work, this call is almost an alarm in view of a clear and present danger.

1.3. A new convention as a solution?

The pursuit of a common legal instrument is maintained in accordance with earlier recommendation from 1982 and 1986, now specified as a European Convention on biomedicine and human biotechnology. In 1991, this proposal becomes the subject of a recommendation on the preparation of a convention on bio*ethics*. This shift in perspective from the techniques of biotechnology to the discipline evaluating them is required, because "combined applications of biology, biochemistry, and medicine, create universal problems which require solutions and have given rise to a new

---

[125]Ibid., 25.
[126]Ibid., 29.
[127]Ibid., 26.

discipline called bioethics. The hopes raised by progress (...) are sometimes tempered by anxiety over the most basic rights of the human person."[128] The idea is to have a convention with general principles beyond the lowest common denominator and additional protocols covering specific aspects. Since the assembly intends to defend basic human rights with a legal instrument containing principles derived from ethical reflection, it bases law on ethical grounds. In the convention, this recommendation is quoted as the most important step from the assembly on this matter.[129]

Considering the developments in biotechnology and the consequences for agriculture, the assembly recommends the inclusion of this field in the bioethical reflection in 1993. Biotechnology is still regarded as ambivalent, offering "important new development perspectives for plant and animal breeding, for the production of food as well as non-food products (energy, pharmaceuticals, medicine) [,]" yet also possible misuse in form of "the production of new diseases or (...) the creation of animals or plants which could have unwanted negative effects on specific ecosystems. The altering of genes and cells and the manipulation of life processes of animals can also result in unnecessary suffering and thus violate animal welfare regulations."[130] The contrast with the enthusiasm originally ignited by the new technology in e.g. 1982 is striking, where the potential for the agricultural sector was thought to "help solve world problems for food production."[131] In 1993, the potentiality of biotechnology is rather seen as limited and aimed at specific, albeit "contrasting aims", spanning from the upgrade of pedigree or indigenous livestock, plants, and food flavour over replacement of chemical herbi- and insecticides to saving endangered species or genetic diversity through embryo banks.[132]

A reluctant tone is also prevailing in a recommendation dealing with the protection and patentability of material of human origin in 1994. The opening line clearly indicates an increasing awareness of human vulnerability: "The Assembly *insists* (my

---

[128]Parliamentary Assembly: Recommendation 1160 (1991), 33.

[129]This is done by way of traditional understatement: "Taking account of the work of the parliamentary assembly in this field, including Recommendation 1160 (1991) on the preparation of a convention on bioethics;" COE: Convention for the Protection of Human Rights and Dignity of the Human Being with regard to the Application of Biology and Medicine: Convention on Human Rights and Biomedicine. European Treaties Series (ETS) 164, preamble.

[130]Parliamentary Assembly: Recommendation 1213 (1993), 36.

[131]Parliamentary Assembly: Recommendation 934 (1982), 15.

[132]Cf. Parliamentary Assembly: Recommendation 1213 (1993), 36.

italics, LR) that human beings are subjects - not objects - of law, that the human body is inviolable and inalienable by virtue of its relationship to a person endowed with rights and that limits must therefore be set to how it is used."[133] Hence, law is subordinated to the human subject, which here is defined in terms of its personhood, possessing a body whose integrity is protected by virtue of that person's inherent rights. Accordingly, the body is regarded as a property, which in turn assumes that the human person is an entity somewhat different from its physical makeup. It is a view which, perhaps, has to link personhood with rationality, since the body is seen as an insufficient criterion for establishing it. After all, its protection is rooted in its 'relationship' to the (human) person, which means that foremost personhood is not understood in terms of its embodiment, but of something else that remains unexplained. Once more, then, the precise nature of human existence seems to elude political discussion and legal definition. We shall return to the roots and bearings of this particular view on human personhood in chapter 2.

Aware of the "rapid development of genetics and the striking range of its present and potential applications [,]"[134] the assembly sees the necessity of protecting investments in the field and, thus, the inevitability of patent laws. At the same time, a fundamental debate on biotechnology cannot be restricted to a mere legal discussion in this regard. Instead, a deep reflection on the human body and the prohibitions and limitations of its commercialisation is implicitly requested.[135] In a remarkably direct manner, the assembly interprets initiatives of the EU with regards to the patentability of living materials as quite insufficient, esteeming that "the approach chosen is simplistic, as much because of the European Union's substantive competence as because its action is geared to the harmonisation of the Single Market and development of Europe's competitiveness and trade."[136] This somewhat enigmatic wording, giving and taking at the same time in a nutshell presents the difference between EU policy and COE as it is depicted by the assembly: the EU as an albeit excellent mechanism for fostering business, yet rather limited when having to grasp the complex consequences of new techniques that may result e.g. in the commercialisation of living materials.

---

[133]Parliamentary Assembly: Recommendation 1240 (1994), 37.

[134]Ibid.

[135]Ibid. The assembly points out that the European Patent Convention, which had been signed prior to the first child born after In vitro fertilisation, was not drafted with such a reflection.

[136]Ibid., 38.

In order to strengthen the role of ethical principles in research and development of living material, then, the assembly reiterates its wish that the 'bioethics convention' be adopted as soon as possible. Furthermore, it recommends drafting protocols which would limit genetic manipulation to human beings and prohibiting various procedures involving the human being, such as cloning, alteration of its genetic identity or the transfer of embryos to another species.[137] As a security measure, the assembly also would like to receive annual reports from the European Patents Office on patents relating to living material.[138]

1.4. Recent initiatives

In 1999[139], progress in transplantation technology had made it possible to speak of a "radical breakthrough" once again, this time "for the transplantation of animal cells, tissues, and organs into humans (xenotransplantation)."[140] Yet, in view of "uncontrollable risks" and "considerable scientific, medical, ethical, social and legal problems", the assembly rejects the fascinating by calling "for the rapid introduction in all member states of a legally-binding moratorium on all clinical xenotransplantation [,]"[141] to be further sustained by a second protocol to the convention on human rights and biomedicine, a worldwide legal agreement, and cooperation with the World Health Organisation (WHO).[142] Obviously, biotechnology had advanced quicker than the desired reflection on its adequacy.

Issued in September the same year, the tone of the recommendation on 'biotechnology and intellectual property', i.e. on the role of patents, is in its specific provisions unconventionally crisp, if not stern. In the introductory parts, the patent system is seen as "an integral part of the market economy and therefore (...) a driving

---

[137]Ibid.

[138]Ibid., 39.

[139]In 1996 and 1997, the assembly voiced its opinion on the drafts of the later convention on biomedicine and its protocol prohibiting human cloning. Since its recommendations were incorporated into these documents, I shall not discuss their contents here, since the exegesis of these texts would warrant an independent study.

[140]Parliamentary Assembly: Recommendation 1399 (1999), 45.

[141]Ibid.

[142]Ibid.

force for innovation."[143] This is hardly contestable, for patents are, of course, a way of securing intellectual property as well as re-compensation and even profit. In this regard, it is interesting to note that the EU increasingly measures the success of developments in biotechnology according to the number of patents filed.[144] According to the assembly, the interests of public lie in the areas of "public order, morality and general aspects of the state economy [,]"[145] the latter indicating a notion of economy as governed rather than liberalised. Hence, it seems that a balance between private and public interests ought to be found. Yet, when dealing with the first specific question of 'living organisms', the assembly declares it "nearly impossible"[146] to find such a balance, since these organisms can reproduce themselves even though they are patented, making it difficult to define the scope of the patent. In the interest of developing countries, farmers, regional and worldwide genetic resources, and the "common heritage of mankind", the assembly "believes that neither plant-, animal-, nor human-derived genes, cells, tissues or organs can be considered as inventions, nor be subject to monopolies granted by patents."[147] Instead, it recommends that the committee of ministers, the EU, the World Intellectual Property Organisation, the Food and Agriculture Organisation, the World Trade Organisation, and Unesco design a suitable alternative system.

A main reference in the document is the Convention on Biological Diversity, signed 1992 in Rio de Janeiro, safeguarding both free scientific access to worldwide genetic resources and the right of developing countries to share the benefits of technological progress. The assembly deems it necessary to "oblige scientists, as well as scientific research and development units working in the field of biotechnology, to conform"[148] with this convention. It also "asks the World Trade Organisation to

---

[143]Parliamentary Assembly: Recommendation 1425 (1999), 51.

[144]Cf. e.g. Phillippe Busquin, European Commissioner for research: Speech at the Belgian-Danish Forum for Innovation in Biotechnology. Brussel 29 May 2002. Unpublished manuscript, 2: "Danish and Belgian researchers and companies are well placed in the European biotechnology league. In the period of 1995-2000, the number of biotechnology patents filed per million inhabitants was highest in Denmark - 4 times higher [than] the EU average. Belgium was third."

[145]Parliamentary Assembly: Recommendation 1425 (1999), 51.

[146]Ibid.

[147]Ibid., 52.

[148]Ibid., 51.

comply"[149] with this convention. In order to demonstrate its attention to the matter, the assembly adds for good measure that it has taken note of the EU directive 98/44/EC on the legal protection of biotechnological inventions from 1998, of the subsequent legal battle, in which the Netherlands and Italy challenged this directive, and of the fact that Norway even "is considering not implementing it."[150]

In the recommendation on biotechnologies issued in 2000, the assembly openly sides with this challenge, calling on the member states of the European Union (!) to request the renegotiation of this directive and to support member governments who have already brought appeals against it before its Court of Justice.[151] In this recommendation, the principle of a healthy environment complements for the first time that of human dignity. Additionally, it "is increasingly important to include ethical considerations centred in humankind, society and the environment in deliberations regarding developments on biotechnologies, life sciences and technologies and their applications."[152] Speaking of "application" in this regard demonstrates that the assembly now uses the term 'biotechnology' primarily to name the notion of an independent entity rather than merely a group of certain techniques. This interpretation is further supported by the choice of title, 'biotechnologies', marking an independent field of diverse practises, which also happens to be in line with the view presented in this book. Already in the introduction, repeating a point made in earlier years, it acknowledges that the "discovery that DNA molecules are interchangeable (...) and the possibility to manipulate or change their units (genes) have given biotechnology enormous scope for application." Thus, biotechnology is again pictured as a recipient of knowledge, storing it for future application, which is expressed by the term "scope." Even though biotechnology appears in the active mode, the agent of application is the individual subject, and the 'scope' is, strictly speaking, merely contained in the mind of a person relating a thing or act to a particular perception.

This development is seen as marked by privations. In general the "[p]ublic opinion should be more strongly involved in political decision-making as regards scientific and

---

[149]Ibid.

[150]Ibid., 51-2.

[151]Parliamentary Assembly: Recommendation 1468 (2000), 54. This highly unusual crossing into the domain of the EU could give the impression that some COE parliamentarians would use the assembly, formed by members of national parliaments, as a platform for EU politics, which of course is very unlikely.

[152]Ibid., 53.

technological choices and scientists should be encouraged to engage more in public debate."[153] As measures of counterbalance, the assembly recommends the "precautionary principle as a common tenet of decision-making", "an assessment method for ascertaining whether new technologies in medicine and biology are compatible with fundamental ethical principles, human rights and human dignity", and a global convention "on the use of living matter."[154] Clearly, the assembly wishes to stress the role of ethical reflection over against the promotion of technology.

In 1999, the assembly also adopted a recommendation on the rights and dignity of the terminally ill and dying. The topic of this recommendation had by some been envisioned as worthy for an additional protocol to the convention on biomedicine. In the course of the political deliberations, however, it became clear that an agreement in this regard was unobtainable, while a mere recommendation was feasible, as its adoption demonstrated. The main reason for disagreement is found at the very end of the document, where it is stressed that "a terminally ill or dying person's wish to die never constitutes any legal claim to die a the hand of another person [and] (...) cannot of itself constitute a legal justification to carry out actions intended to bring about death."[155] In other words, the legal right to euthanasia, i.e. killing on request, is rebuked, which collides a little with the provisions or applications of law in some European countries.[156] Euthanasia is rejected, because it is regarded as one practice among others that threaten the dignity of the terminally ill or dying person, which is the main criterion referred to in the document. Since the assembly wishes to protect the right to self-determination, however, palliative care that might hasten death is explicitly accepted, as are advance directives, living wills and the desire to discontinue or refuse

---

[153]Ibid.

[154]Ibid., 53-4.

[155]Parliamentary Assembly: Recommendation 1418 (1999), 50.

[156]In its reply to this recommendation, the committee of ministers "notes that the legal position differs from one member state to another on advance refusal of certain treatments and on euthanasia." Parliamentary Assembly of the Council of Europe: Protection of the human rights and dignity of the terminally ill and the dying. Doc. 8888 7 November 2000. Recommendation 1418 (1999). Reply from the Committee of Ministers adopted at the 728th meeting of the Ministers' Deputies (30 October 2000). In the same document, when referring to the human right of self-determination as expressed in article 9 of the "Bioethics Convention", the committee thereby uses a name for the convention, which was not adopted after all.

medical treatment.[157] In a comprehensive list, the assembly recommends a variety of measures that are to ensure adequate care in the broadest sense, ranging from palliative care to the training of professionals and support of proxies assisting humans facing death.[158]

As the committee of ministers underline in their reply to this recommendation, the assembly, thus, wishes to respect and protect the dignity of this group of persons in threefold way: "stressing, first, access to care, including palliative care; second, the terminally ill or dying person's right to self-determination; and, third, the prohibition on intentionally taking the life of a terminally ill or dying person."[159]

The committee gave another, more specific reply in 2002, relating these main issues "to the one incontestable area of Council of Europe competence: human rights protection under the European Convention on Human Rights and the case law of the European Court of Human Rights."[160] It concludes that the *only* derogations from the right to life are those mentioned in article 2 of the ECHR, namely the death penalty, self-defence and killing in action, among which euthanasia consequently is not.[161]

---

[157]Parliamentary Assembly: Recommendation 1418 (1999), 49.

[158]Ibid.

[159]Parliamentary Assembly of the Council of Europe: Protection of the human rights and dignity of the terminally ill and the dying. Doc. 8888 7 November 2000. Recommendation 1418 (1999). Reply from the Committee of Ministers adopted at the 728th meeting of the Ministers' Deputies (30 October 2000).

[160]Parliamentary Assembly of the Council of Europe: Protection of the human rights and dignity of the terminally ill and the dying. Recommendation 1418 (1999). Doc. 9404 8 April 2002. Reply from the Committee of Ministers adopted at the 790th meeting of the Ministers' Deputies (26 March 2002), 1.

[161]In the ECHR, article 2 states: "1. Everyone's right to life shall be protected by law. No one shall be deprived of his life intentionally save in the execution of a sentence of a court following his conviction of a crime for which this penalty is provided by law. 2. Deprivation of life shall not be regarded as inflicted in contravention of this article when it results from the use of force which is no more than absolutely necessary: 1. in defence of any person from unlawful violence; 2. in order to effect a lawful arrest or to prevent the escape of a person lawfully detained; 3. in action lawfully taken for the purpose of quelling a riot or insurrection." The European Court of Human Rights has ruled that this article "not only safeguards the right to life, but sets on the circumstances when the deprivation of life may be justified; Article 2 ranks as one of the most fundamental provisions in the Convention - indeed one which, in peacetime, admits of *no* (my italics, LR) derogation under Article 15. Together with Article 3 of the Convention, it also enshrines one of the basic values of the democratic societies making up the Council of Europe. As such, its provisions must be strictly construed." European Court of Human Rights, McCann and others v. the United Kingdom, 27 September 1995, § 147.

Furthermore, relieving human suffering is seen as contributing to the protection of human rights and dignity and, therefore, to be supported. At the same time, there are questions of how to solve possible conflicts of rights and freedoms, of the precise impact of international law on domestic law and of whether the convention guarantees negative rights and whether an individual can renounce the exercise of certain rights.[162] This enterprise is not eased by the need for further clarification of what precisely constitutes and characterises notions such as 'dignity', 'suffering' and 'adequate care'. Of course, their semantic imprecision facilitates agreeing on their undoubtable worth, for they are very much open to individual interpretation.

Almost twenty years after its 1982 recommendation on genetic engineering, the assembly also returns to some of the language used then in its recommendation on the protection of the human genome by the council of Europe, adopted in 2001.[163] The first draft of the human genome had been presented and again, the assembly was torn between utmost fascination and slight rejection, expressed as caution. An example of the first is found in the introduction: "The genetic age will dawn with the completion of the project: diagnosis will become objective, and it will be possible to identify (...) genetic disorders or a genetic predisposition to illnesses at an early stage. In many cases, gene therapy will become possible, and this will basically give rise to a form of genetic engineering designed, for instance, to avoid the development of tumour in an individual found to be at risk. It might also be applied to other illnesses, such as hypertension, diabetes, Alzheimer's disease, osteoporosis, certain psychiatric disorders, etc."[164] It acclaims its promise of "enormous improvements in the quality of life", the "incalculable opportunities for preventing illness and improving treatment [,]" and also, a little sourly, the "grandiose research effort - in which the United States has the lead

---

[162]These points are expressed in much more careful legal terms, since they pertain to the very delicate issue of how COE regulation affects state law. The delicacy lies in the fact that all COE legal instruments only are in force if a state has agreed to it by signing and ratifying them. Having thus limited its sovereignty, the question is whether acceptance can be withdrawn. This is another way of phrasing a central human problem, namely whether humans can renounce their autonomy, in particular in terms of their freedom. Parliamentary Assembly of the Council of Europe: Protection of the human rights and dignity of the terminally ill and the dying. Doc. 8888 7 November 2000. Recommendation 1418 (1999). Reply from the Committee of Ministers adopted at the 728th meeting of the Ministers' Deputies (30 October 2000), 3-4.

[163]Parliamentary Assembly: Recommendation 1512 (2001), 55.

[164]Ibid.

over Europe".[165]

Having learned from the previous development, however, the assembly wishes a body to monitor "the development of the Human Genome Project research process", including research site visits, mandatory consultation of the European authority, national boards of supervision, signature and ratification of the convention on biomedicine and human rights, and the change of patent law "as far as the ownership of human being tissue and genes is concerned, into law pertaining to the common heritage of mankind."[166] Hence, technological advance requires ethical scrutiny. This position is taken in order to protect human dignity and the respect for life, those "universally recognised ethical and moral principles"[167], and instigated by the ethical problems pertaining to cloning cells, to the definition of the conditions for genetic testing, especially with regards to insurance and employment, and to the assessment of the health risks related to such research.[168]

1.5. Results

In its discussions on biotechnology, the assembly shows a gradually sharpened awareness of possible negative effects caused by the new techniques. It does not reject biotechnology as a project, still considering it a motor of progress, but there is an increasing fear to be sensed with regards to possible inflictions on the dignity and rights of the human being as well as on the fabric of society and, most recently, on the environment.

As an effective means of counterbalance, the assembly has since 1982 envisioned a common European initiative in the field. In section 3.3., we shall discuss how the European convention on biomedicine, opened for signature 4th April 1997, materialises this vision.

Meanwhile, the assembly has begun to propose initiatives beyond the boundaries of Europe in order to control the biotechnological entity universally, and its requests and recommendations, feeble instruments as they may be, demonstrate the will to base technology on values rather than mere momentum.

---

[165]Ibid., 55-56.
[166]Ibid., 57.
[167]Ibid., 56
[168]Cf. ibid., 55-6.

2. THE COMMITTEE OF MINISTERS

In view of the developments in biotechnology, the committee decided in 1985 to set up an ad hoc committee of Experts on Biotethics (CAHBI) under direct authority of the ministers, which in 1992 became the steering committee on bioethics (CDBI), "responsible for the intergovernmental activities of the Council of Europe in the field of bioethics."[169] The significance of this group can be measured by the number of specific recommendations and protocols issued and last not least by the succesful drafting of the convention on biomedicine.

2.1. Systematic beginnings

Since its 1978 resolution on harmonisation of member states relating to removal, grafting and transplantation of human substances, the committee of ministers has issued recommendations with rules related to biotechnology in the broadest sense, i.e. also covering issues in biomedicine: transplantation, persons suffering from mental disorder, the use of DNA for experiments and within the criminal justice system, medical research on humans, genetic screening and counselling, the use of human blood and tissue, the protection of medical data, medico-legal autopsy, and xenotransplantation. These set of rules are intended as models for national legislation aimed at European harmonisation and typically structured in a legal manner using articles or principles.

   In the resolution, the committee excludes explicitly "the transfer of embryos, the removal and transplantation of testicles and ovaries and utilisation of ova and sperm"[170], without defining e.g. the status of the embryo. From the rules of this resolution, four principles can be extracted, which with minor refinements have proven quite robust in the subsequent committee work:
- informed consent of persons concerned
- without explicit consent, procedures require therapeutic or diagnostic rationale
- procedures should cause the least possible risk
- substances may not be offered for profit.

The principle of consent rests fundamentally on the assumption of human freedom as presented above: since free, humans are entitled to act freely. Consent is in this regard

---

[169]Zilgalvis (2002), 2.
[170]Council of Europe Committee of Ministers: Resolution (78) 29, 47.

an expression of freedom, namely the freedom to subject oneself. It also presupposes the human ability to communicate and to fathom what is being communicated, interacting in the common sphere of human rationality with its means of language, thought and will.

The principle of the least risk also presupposes freedom, namely the freedom of choice, the ability to assess the risks at stake, i.e. to anticipate the consequences of an act, in addition to maintaining the idea of preserving the integrity of the human body by harming it as little as possible.

The reluctance to allow profit as a permissible motive for action stems from the fear that it may influence one's freedom unduly to accept risks one would otherwise not, assuming that when free, any human would choose what is less risky. In short, one may see these principles as reflections of human freedom and the conditions ensuring its exercise.

In 1979, the committee notes "the increasing demand for human substances" and adheres that it "must be further facilitated by common action of the member states in order to make them available in due time and condition." This demand is nourished by the "substantial increase (...) in the treatment of patients by transplantation or grafting of collected human organs, tissues or other substances (...)"[171] Thus emerges the chain of biotechnology: having identified problems (e.g. a 'disease'), research crafts techniques (e.g. a 'treatment') requiring substances (e.g. a 'human organ') provided by either a donor (e.g. by 'transplantation') or engineered e.g. genetically, which requires research anew.

At that time, working with DNA was still at an early stage, and following the recommendation on genetic engineering given by the assembly in 1982 and the EEC recommendation of the same year[172], the committee issued its recommendation on the subject in 1984, focussing on the necessity of notification in conjunction with this work, defined as "the formation of new combinations of genetic material by the insertion of nucleic acid molecules produced by whatever means outside the cell, into any virus, bacterial plasmid or other vector system so as to allow their incorporation into a host organism in which they do not naturally occur but in which they are capable of continued propagation."[173]

---

[171]Council of Europe Committee of Ministers: Recommendation R (79) 5, 51.

[172]European Communities Council Recommendation of 30 June 1982 (82/472/EEC).

[173]Council of Europe Committee of Ministers: Recommendation R (84) 16, 59.

The committee "welcomes" the improved safety in the field, but is also aware of possible biohazards caused by it, to which the notification of research projects and their accompanying security measures are envisaged as a safeguard. Owing to the character of the intergovernmental structure of the COE, however, the committee has to let the national states define the categories of such hazard. The system of notification envisioned is, thus, dependent upon the anteceding national definition of the instances to which it will apply. This is quintessentially the dilemma of any COE regulation, for even the ostensibly clear values of human dignity and human personhood need definition and contextualisation, which in turn endanger the absolute protection they allegedly enjoy.

2.2. The struggle for standards

With the 1990's, the struggle for identifying common standards for the new techniques is increasingly evident. Human dignity is identified as *the* criterion for evaluating medical research on human beings in the recommendation of 1990, where the committee is convinced "that medical research should never be carried out contrary to human dignity[.]"[174] The recommendation is issued on the assumption that this research should consider ethical principles and be subject to legal provisions. The medical sector is, thus, not seen as an independent and neutral entity.

The four principles formulated earlier reappear, one of which in refined form. Almost identical are the provisions demanding only minimal risk of such research; the free, informed, express and specific consent of the person undergoing it; and the absence of financial inducements compromising that consent.[175] With regards to the legally incapacitated, i.e. those with an assumed inability to consent, the committee stresses that these persons and those deprived of liberty may only undergo research of direct and significant benefit to their health, which substitutes the former stipulation of "therapeutic and diagnostic reasons." This principle is less strictly applied to pregnant or nursing women or children, on whom research also may be carried out even if it is of no direct benefit to them, provided it is beneficial to "other women and children who are in the same position and the same scientific results cannot be obtained by research

---

[174]Council of Europe Committee of Ministers: Recommendation R (90) 3,, 60.
[175]Ibid., 60-64, principles 2, 3 and 13.

on women who are not pregnant or nursing."[176]

While still open to interpretation, the stipulation of "direct benefit" is at least less vague than "therapeutic or diagnostic reasons", since it links the research to the benefit of the actual person concerned. Since the protection of the individual already in the beginning of the document is announced as the "paramount concern[,]"[177] the introduction of this principle can be interpreted as a reinforcement of this endeavour. Insisting that "the interests and well-being of the person undergoing medical research must always prevail over the interests of science and society[,]" the committee introduces a principle against the weighing of interests in this regard. Aware of the "considerable progress (...) achieved in detecting genetic abnormalities in the child to be born through genetic screening and through parental diagnosis of pregnant women," the committee sets forth a recommendation for this field in 1990, "noting the fears that these procedures arouse."[178] Once more, the committee acknowledges progress in a field of biotechnology, while realising its concomitant risk of offending "the moral values which are the common heritage of the member states [.]"

At this point, the committee presents a sketch of its value system: since the moral values are "based essentially" on the respect for life and human dignity, "life" and "dignity" are the most fundamental values, on which the so-called "moral values" and its subsequent "ethical, medical, legal and social principles" are based. In this way, "life" and "dignity" are understood as values with inherent qualities, the respect of which is rooted in the respective value itself and not in our acceptance of it. They are, thus, intrinsic values. Furthermore, it seems that the term "moral values" in the present context is regarded as a heading for the various values identified as "the common heritage of the member states," implicitly accepting the respect for fundamental values to be expressed differently through the definition of "moral values". It is also interesting that the term "ethics" appears as an adjective used in line with "medical, social, and legal principles [,]" presenting it almost as a professional discipline of application, likewise employed to govern procedures in order to prevent abuse, rather than as an interdisciplinary field of reflection and evaluation. Hence, the principles laid out in the recommendation are not considered as the foundation for, but as an application of moral values based on the two fundamental values of "life" and

---

[176]Ibid., 62.

[177]Ibid., 61.

[178]Council of Europe Committee of Ministers: Recommendation R (90) 13, 65.

"dignity." Thus, the notion of 'principles' here equals that of 'working tools.'

"Prenatal genetic screening" is defined as "screening tests carried out to identify among the general population of apparently healthy individuals those at risk of transmitting a genetic disorder to their offspring." The concept of 'disease' is applied in a very broad sense, for even the mere risk of transmitting a "genetic disorder" to one's offspring makes one "apparently healthy", i.e. not healthy and, thus, ill. Also, "genetic disorder" is not defined, yet with principle 14 suggesting preconception counselling "when there is an increased risk of passing on a serious genetic disorder," the committee at least operates with the two categories of serious and non-serious genetic disorders. The various forms of screening are envisioned as means of detecting actual or possible forms of health risks, embedded in "non-directive"[179] counselling and requiring "free and informed consent of the person concerned."[180] This provision on prenatal genetic screening is a little imprecise, for one might consider the embryo or foetus quite as much concerned as the woman, unless neither embryo nor foetus are regarded as *persons*, which brings us back to the obvious lack of definition in terms of human personhood and being. In this regard, it is uncertain whether the protection of "the woman's freedom of choice"[181] comprises more than accepting or refusing screening and diagnosis.

In general, the precise scope of this recommendation remains opaque: while supporting various forms of genetic screening as optional health care procedures, the recommendation does not elaborate on the measures to be taken after "a serious risk to the health of the child"[182] has been detected.

In view of the "development of the biomedical sciences[,]" and acting on a proposal submitted by the COE Secretary General, the European ministers of justice decided the same year to initiate the process leading to a "framework convention, open to non-member States, setting out common general standards for the protection of the human person in the context of the development of the biomedical sciences[.]"[183] The resolution reaffirms central features of the COE value system: its commitment to the principles of human rights, human dignity and the rule of law; its wish to protect the

---

[179]Ibid., 67, principle 4.

[180]Ibid., principle 6.

[181]Ibid., principle 9.

[182]Ibid., principle 2.

[183]17th Conference of European Ministers of Justice (1990): Resolution No. 3.

fundamental rights of human beings affected by this development; and the need to protect these rights internationally. The document also contains a semantic imprecision, speaking both of the fundamental rights of human beings and the universality of the rights of the human *person*.[184] Imprecision in this regard would also mark the later convention.

Having applauded the use of DNA analysis in the criminal justice system in another recommendation, "[c]onsidering that the fight against crime calls for the use of the most modern and effective methods [,]"[185] the committee returns to the topic of genetic screening in a 1992 recommendation. Again, "recent progress in the field of biomedical science has made it possible to obtain a greater knowledge of the human genome and the nature of genetic disorders [,]"[186] shaping the techniques of genetic testing and screening, which "can be carried out at different levels, such as on chromosomes, genes (DNA), proteins, organs or a given individual, and can be complemented with aspects of the family history."[187] The committee also understands the anxiety aroused by these techniques and is determined to fight the phenomena of "discrimination and social stigmatisation, which may result from genetic information [.]"[188] Defining "genetic tests for health care purposes[,]" the role of the test is e.g. envisioned "to diagnose and classify a genetic disease [.]"[189] As discussed earlier, using genetic testing in this regard implies the setting of an examiner employing a particular notion of disease in his analysis of the object.

This act of classification is a means of understanding, comparing an actual instance with a model extrapolated from previous experience. Such abstractions are a little awkward with regards to genetic 'diseases,' because they often are predispositions rather than manifestations. Moreover, the dispositions will actualise themselves with varying degrees of likelihood, ranging from possibility over probability to certainty.

Governing principles in the document are the equality of access to testing and self-determination through express, free and informed consent. In application of these principles, compulsory tests are seen critically, yet accepted when expressly allowed

---

[184]Ibid.

[185]Council of Europe Committee of Ministers: Recommendation R (92) 1, 69.

[186]Council of Europe Committee of Ministers: Recommendation R (92) 3, 74.

[187]Ibid., 75, footnote 1.

[188]Ibid., 75.

[189]Ibid.

by law "for the protection of individuals or the public."

Consequently, while it is stated that "insurers should not have the right" to require genetic testing or results of any previously undertaken, this provision could be overruled by national law. It is not clear, then, how strong the protection of the subject indeed is, in particular because the scope of any recommendation is limited by the facts of national law. This insecurity is further nourished by the vagueness of terms open to interpretation, e.g. when declaring that genetic data should "*as a general rule* (my italics, LR) be kept separate from other personal records."[190]

In 1993, the "growing importance of blood products in supportive haemotherapy and the need to subject such products to clinical testing and trials" led to a recommendation on products derived from human blood or plasma. While the procedures designed to ensure "their safety, efficacy and quality" are comparable to other medicinal products, the human origin of the products necessitates taking "specific ethical and technical principles" into account "in addition to those, national and international applying to medical research and clinical trials on human beings."[191] The committee presents the ethical principles of consent, non-enumeration and the good health of the donor in order to protect donor and recipient "against transmission of diseases or against medicinal products and drugs which could be detrimental to him/her."[192]

The technical principles focus e.g. on good manufacturing practice and the permissibility of randomised and double blind trials. They also include ethical reflections on the use of placebos in "clinical trials of blood components and fractionated products[,]" which is considered problematic, "since it is unethical to withhold treatment." As a placebo is not a treatment, but a means faking it, its administration can be regarded as withholding treatment for someone in need of it.[193]

---

[190]Ibid., 79.

[191]Council of Europe Committee of Ministers: Recommendation R (93) 4, 80.

[192]Ibid., 82.

[193]Weijer (2002) considers the use of placebos problematic, because it is carried out with a strong scientistic bias: "Whether a placebo-controlled trial can assess the contribution of various treatment components rests on the assumption that each treatment component has an additive effect. Only on this additive model can the observed effect be thought to be constructed from a series of components, each adding a discrete proportion of the observed efficacy." (70) This view is too simplistic, opines Weijer, because psychological and physical effects may interact. Indeed, "the clinician's office is not a laboratory anymore than a research subject is a lab rat. Clinical care and human response to disease

Physical and chemical modification of blood components need approval by an ethical committee, "unless the changes are such that secure in vitro tests demonstrate that there has been no biological change."[194] Using biological integrity as a benchmark, the committee indirectly refers to the value of life and the respect due to it.

The recommendation contains a glossary of the central terms used, including a description of the four phases of clinical trials for the development of medicinal products. It is mainly a document intended to standardise the clinical use of human blood components, while stressing the duty of respecting the special character of human material. Although much briefer and less technically detailed, the recommendation on human tissue banks following in 1994 has quite similar objectives, incorporating the same basic principles as the preceding documents.[195] Defining human tissue, the committee includes all "constituent parts of the human body[,]" but excludes organs, blood and "reproductive tissue, such as sperm, eggs and embryos." Thus, the human embryo is now defined as "reproductive tissue," yet still regarded as a constituent part of the human body, even though it is excluded from the scope of the recommendation.

Two months before the convention on human rights and biomedicine was opened for signature, the committee launched a comprehensive recommendation on the protection of medical data, replacing an earlier from 1981. It stresses the rights and fundamental freedoms of the individual "and in particular the right to privacy." The matter of defining the human being is muddled by introducing a new category, namely the "unborn child", which is not defined either. In parts, the recommendation mirrors provisions to follow in the convention, e.g. every person's right to access his or her medical data.[196]

In view of the risk of transmitting disease through xenotransplantation, which "may become a practicable therapeutic intervention in the very near future [,]" the committee recommends establishing a mechanism for the registration and regulation of this field in 1997.[197]

---

are simply too complex to be captured by such a simplistic model." (71)

[194]Council of Europe Committee of Ministers: Recommendation R (93) 4, 83.

[195]Council of Europe Committee of Ministers: Recommendation R (94) 1.

[196]Council of Europe Committee of Ministers: Recommendation R (97) 5, 95.

[197]Council of Europe Committee of Ministers: Recommendation R (97) 15, 100.

2.4. Results

The work of the committee reflects the wish to protect human dignity and the right to life expressed through a variety of principles intended to secure the expression of human freedom and the conditions for its use. The tones of fascination and rejection are less strongly voiced than in the assembly, which also may reflect the different setup of the committee as an executive committee of government representatives. It operates with a value system, consisting of fundamental values and their inherent rights, secondary principles derived from them and positive law employed at the national and international level as a means of safeguarding the values of common European heritage. The precise nature of the underlying values is not examined and especially in terms of defining the human personhood, the committee does not arrive at taking a clear stance, which is unfortunate, since the human being is the agent as well as often the object of the techniques forming biotechnology. Still, in its dealings with this field, the committee proceeds cautiously, seeking to balance the effects of the new techniques mainly with the instruments of law and the institutions supervising its rule, nicely reflecting the commission of the COE.

3. COUNCIL OF EUROPE CONVENTIONS

Deliberations on biotechnology within the COE have also resulted in the convention on biomedicine[198] with its protocols on cloning[199], organ and tissue transplantation[200] and further "draft additional Protocols on biomedical research, protection of the foetus and the human embryo, and on human genetics [...] currently being drafted by working parties made up of high level experts nominated by the Council of Europe member States with the assistance of the Secretariat of the Council of Europe (the Bioethics

---

[198]Council of Europe: Convention on Human Rights and Biomedicine. By July 2002, the convention has been signed by 31 states and ratified by Cyprus, Czech Republic, Denmark, Estonia, Georgia, Greece, Hungary, Portugal, Romania, San Marino, Slovakia, Slovenia and Spain, where it is now in force.

[199]Additional Protocol to the Convention on Human Rights and Biomedicine on the Prohibition of Cloning Human Beings. Paris 12 January 1998. European Treaty Series 168.

[200]Additional Protocol to the Convention on Human Rights and Biomedicine, on Transplantation of Organs and Tissues of Human Origin. Strasbourg 24 January 2002. European Treaty Series 186.

Section in the Directorate of Legal Affairs)."[201]

As stated by its title, the convention is designed for the protection of human rights and dignity of the human being with regard to the application of biology and medicine. Hence, it is an instrument created to prevent the "accelerating developments in biology and medicine"[202] from harming the human being by protecting its dignity, identity, integrity, interests, welfare, fundamental rights and freedoms. It is, thus, envisioned as a particular means of protection within the larger framework of international legal instruments considered relevant in this respect by the COE.[203]

3.1. The European Convention on Human Rights

The European convention for the protection of human rights and fundamental freedoms (ECHR), opened for signature in Rome on 4 November 1950, and most recently amended by Protocol No. 11,[204] declares in article 2, 1: "Everyone's right to life shall be protected by law. No one shall be deprived of his life intentionally save in the execution of a sentence of a court following his conviction of a crime for which this penalty is provided by law." Without engaging in the complex discussion of the bearings of this article, it is in our context relevant to point out that the life of the individual is considered a fundamental value, anteceding law and warranting a legal right only to be overruled in case of the death penalty, in conjunction with a lawful arrest or when settling public insurrection. In addition to these exceptional circumstances, personal self-defence against an attacker is not seen as an infliction on this provision, either. With law and its execution superceding an individual right inherently given whenever someone is alive, though, law claims authority even over the values it is erected upon. The same approach is pursued for the rights of liberty, security, privacy, and family life, also guaranteed in articles 5 and 8, yet likewise to be

---

[201] Zilgalvis (2002), 3.

[202] Council of Europe: Convention on Human Rights and Biomedicine, preamble.

[203] In the preamble of the convention, the most relevant instruments are listed: the UN Universal Declaration of Human Rights (UNDHR); the (European) Convention for the Protection of Human Rights and Fundamental Freedoms (ECHR); the European Social Charter; the International Covenant on Civil and Political Rights, the International Covenant on Economic, Social and Cultural Rights; the Convention for Protection of Individuals with regard to Automatic Processing of Personal Data; and the Convention on the Rights of the Child.

[204] Council of Europe: Convention for the Protection of Human Rights and Fundamental Freedoms as amended by Protocol No. 11, Rome, 4.XI.1950. European Treaties Series (ETS) 5.

overruled by law e.g. "for the prevention of the spreading of infectious diseases, of persons of unsound mind, alcoholics or drug addicts or vagrants"[205] or, more vaguely, "for the protection of health or morals[.]"[206] This last stipulation is also used as a possible limitation on the freedom of thought, conscience and religion (art. 9), the freedoms of expression (article 10) and of assembly and association (art. 11).

This notion of health reflects a utilitarian position in subjecting the individual to the standards set by others through their interpretation of the very general and, thus, imprecise terms "health", "morals", "public security", "lawfulness" etc.

3.2. The European Social Charter

Complementing the ECHR in the dimension of social rights, the European social charter[207] was opened for signature in Turin 18 October 1961. It declares the right to health, more precisely "the right to benefit from any measures enabling him to enjoy the highest possible standard of health attainable," further specified as removing "as far as possible the causes of ill-health[,] to provide advisory and educational facilities for the promotion of health and the encouragement of individual responsibility in matters of health [and] to prevent as far as possible epidemic, endemic and other diseases." Simply put, health is here understood as a variable quality that can be obtained, especially if the causes of its privation disappear and the access is granted to anyone regardless of their resources.[208] In order to achieve this goal, prevention is necessary, which in turn requires promotion and education, also encompassing working conditions.[209]

This endeavour is result of a complex interaction: it assumes a common or at least politically accepted perception of a health standard, of ways of obtaining it for those who are not considered up to it, of the possibility of communicating this ideal and of the will to achieve it. In short, the charter defends the notion of public health. It

---

[205]Article 5

[206]Article 8

[207]European Treaty Series No. 35

[208]According to article 13, the right to social and medical assistance includes adequate assistance given to those under a social security scheme with full respect to their political or social rights.

[209]Article 3 promotes the right to safe and healthy working conditions.

presupposes the ability of individuals to fathom the notion of health as a distinct state of life as opposed to its privations and to agree on a common interpretation, further requiring agents willing to undergo personal change instigated by education and promotion and, thus, surpassing the traditional scope of health care as merely responding to an immediate need. Yet, as seen before, the main problem lies in the definition of terms.

3.3. The European Convention on Human Rights and Biomedicine

The difficulty of defining terms also marks the convention on Human Rights and Biomedicine, which is particularly apparent in its struggle for terminological clarity regarding the human being. A survey of its preamble and first chapter reveals a view, according to which "a being that is human possesses, to the extent that is human, an essential dignity and identity, because it is *human*, i.e. of the human species. At the point of individualization of human nature, the chronology of which is debated and, therefore, left to unspecified domestic law to determine, this human being has become 'one' of 'every*one*' or, analogously, '*une personne*' among '*toute personne*', i.e., an individual human being, which in turn ensures its inherent integrity, demanding respect and, thus, establishing subsequent fundamental rights and freedoms of the individual."[210]

Thus, the human being essentially possesses dignity and identity, while the individual evolving from it enjoys an inherent integrity with subsequent rights and freedoms. The entire chapter two "deals with the consent of 'the person,' thus introducing the notion of personhood that seems to have been avioded in the preamble and the first chapter. In fact, beginning with that chapter, 'person', 'individual', 'everyone', and 'patient' are used synonymously, and they also substitute the term 'human being' without further reference."[211] Human embryos are understood as a different instance of human life, resulting in three different legal categories, namely 'human life', 'embryo' and 'personhood'. The human being can, thus, be an individual and consequently a person, or an embryo, both of which are instances of human life. The notion of rationality is not used as a criterion to determine the precise character of this life, in spite of the dominant role assigned to the principle of consent, demanding

---

[210] Reuter (2000), 186.
[211] Ibid.

that any "intervention in the health field may only be carried out after the person has concerned has given free and informed consent to it."[212] If consent cannot be given, the principle of direct benefit is employed in its place, e.g. in art. 6, allowing for interventions on persons not able to consent for the purpose of research or organ removal, as long as this principle is respected.

According to the explanatory report to the convention, persons may lack this ability "to give full and valid consent to an intervention due to either age (minors) or their mental incapacity."[213] The convention leaves it to domestic law to decide, when a minor or an adult does not have the capacity to consent, respecting the autonomy of national law as well as avoiding to define personhood in terms of potential or actual rationality. The principle of direct benefit is in turn substituted by another principle in article 17. Here, research on persons without the capacity to consent is accepted under certain conditions provided it is intended to benefit the same group to which the person belongs, e.g. in terms of age. The conditions include the lack of alternative, proportionate risks to potential benefits, approval of the project, sufficient information, authorisation in writing and that the person concerned does not object to the procedure.[214]

So, we encounter a degression from the internal criterion of consent, expressed by the subject concerned and legally accepted on the assumption of rational competence, first to the externally determined criterion of direct benefit primarily and then to what might be called 'group benefit'. This creates a tension of which the COE is well aware. In the explanatory report, it defends its stance taken between respecting the individual and ensuring the freedom of research: "Were such research to be banned altogether, progress in the battles to maintain and improve health and to combat diseases only afflicting children, mentally disabled persons or persons suffering from senile dementia, would become impossible."[215]

The use of martial language in this context is quite deliberate, I think. War is usually considered a situation requiring extraordinary measures, setting aside individual

---

[212]Council of Europe: Convention on Human Rights and Biomedicine, article 5.

[213]Council of Europe, Directorate of Legal Affairs: Explanatory Report DIR/JUR (97) 1, point 41.

[214]Art. 16, i-iv and art. 17 i-v.

[215]Council of Europe, Directorate of Legal Affairs: Explanatory Report DIR/JUR (97) 1, point 107.

interests for the sake of the public cause. The quest for health is clearly seen as such a cause, with disease, or more fundamentally, the imperfection of human nature, identified as the enemy to be fought. Health is, thus, placed in the larger context of biotechnology, surpassing merely easing the actual state of a person as in traditional health care, by conscribing a collective enterprise not unlike warfare. Those who become objects of research without having the capacity to consent to it are subjected to this cause with the expectation that its achievement also may benefit them, just as peace after war might do. It is for this reason, then, that the principle of consent is supplanted with principles based upon external evaluation.

It is almost standard practice to use martial language when institutions involved in the promotion of health present strategies for dealing with diseases, typically culminating in the metaphor of "declaring war on disease." For the sake of argument, then, let us refer to a leading theorist of war, namely Carl von Clausewitz. For him, the concept of war exists as pure abstraction, the absolute war created by the escalation of violence, and in a moderate form as real war, "where the irrationality of violence and the force of the military will, are subjected to political rationality [,]"[216] requiring politics, strategy and tactic: Politics addresses ends, means and the people. Strategy is the questions of means, of which tactic is but one.[217]

Applied to the level of the COE, a convention would be a tactic used as part of the strategy to harmonise e.g. health laws in Europe in order to maintain and improve health. Yet is it indeed appropriate to consider illness and disease as an enemy to be conquered? Clearly, the declaration of such war relies on the presence of an enemy, which readdresses the question of interpretation in this way: at what point is a human state of being a disease or an illness, and moreover in need of a public response?

At several points, the document operates with other terms open to interpretation. Article 2 states the primacy of the human being by declaring that its interests and welfare "shall prevail over the *sole* [my italics, LR] interest of society or science." This qualification bears a little resemblance to one of the formulations of Kant's categorical imperative, according to which one should always treat the other not merely as a means, but always also as an end.[218] Acknowledging the legitimacy of interests on both sides, however, does not solve the problem, but merely identifies it. Article 3 seeks to ensure

---

[216] Verstraeten (1985), 56.

[217] I have here used Johan Verstraeten's solid interpretation presented ibid.

[218] Immanuel Kant: Grundlegung zur Metaphysik der Sitten, A 66f.

"equitable access to health care of *appropriate* quality[,]" which leaves ample room for national differences. If a patient has previously expressed wishes relating to medical intervention, these "shall be taken into account" according to article 9, but this warrants neither a right, nor does it give criteria for limitations on this account. Concerning the human genome, article 13 stresses that any modification "may only be undertaken for preventive, diagnostic or therapeutic purposes, and only if the aim is not to introduce any modification in the genome of any descendants." While the second stipulation is fairly concrete, the first grants a certain flexibility, for most research can be related to either of these purposes.

Still, the convention also contains provisions of a remarkable clarity and strength: as mentioned earlier, article 5 requires the consent of the person concerned and allows for the consent to be withdrawn freely "at any time." Article 10 declares in part 2: "Everyone is entitled to know any information collected about his or her health. However, the wishes of the individuals not to be so informed shall be observed." This stipulation is strong enough to have caused a reaction in form of a reservation on the freedom granted. here. Equally strong formulations are the prohibitions of discrimination on grounds of genetic heritage (art. 11) and of the creation of embryos for research purposes (art.18, 2).

Even these inscriptions of fundamental rights, with the exception of articles 11, 13, 14, 16, 17, 19, 20 and 21, are subordinated to the general provision contained in article 26, according to which the exercise of these rights may be restricted when "prescribed by law and (...) necessary in a democratic society in the interest of public safety, for the prevention of crime, for the protection of public health or for the protection of the rights and freedoms of others." The elasticity of these terms is, of course, necessary to accommodate for the national differences in interpreting the notions of e.g. public safety or health, but it also illustrates the dilemma of defining international standards with legal instruments dependent upon national acceptance and implementation.

Concluding, we encounter a convention with tensions stemming from the conflicting interests of protecting the human being against the risks emerging from the developments of biotechnology as well as the freedom of research in the field. This freedom may entail using material of human origin or carrying out research on those who at least legally are considered incompetent to give consent, even though perhaps able to voice their objections. The lack of clear definition of the human being reflects the differences existing between the various nations with regard to the use of e.g. embryos for research, for if the human being is defined as existing from the time of conception on, this would create certain restrictions on the freedom of research. It is telling that Belgium declines signature of the convention, considering the ban on creating embryos for research to strict, while Germany likewise declines, *inter alia* judging the protection of the embryo too weak.[219]

At the very sensitive points of determining the character of human life, the absence of the capacity to consent and the scope of potential benefits, the convention has to let domestic law define the borders, then, turning it into an instrument of stronger intentions than means.

3.4. The Protocol on Human Cloning

The first protocol to the convention, issued in 1998, is a response to the emerging technical possibility of cloning human beings.[220] Apart from stipulations on technicalities, it consists basically of two articles: article 1, 1 prohibits"[a]ny intervention seeking to create a human being genetically identical to another human being, whether living or dead[,]" with "genetically identical" in paragraph 2 defined as "a human being sharing with another the same nuclear gene set." The prohibition is excluded from interpretations of national law, as otherwise provided for by art. 26 in the convention, and based on the following line of argument, presented in the preamble: deliberately creating genetically identical human beings is a misuse of biology and medicine, because as an act of instrumentalisation, it is contrary to human dignity.

This is a petitio principii, since the act can be only said to be against dignity after having been defined as an unethical use of a human being. In addition, the protocol envisions "serious difficulties of a medical, psychological and social nature (...) for all the individuals involved[.]" While acknowledging the possible advantages for "scientific knowledge and its medical application", the COE thus at this point clearly opts for limiting the freedom of research in order to protect the individual.

The notion of "scientific knowledge" supports a point made earlier with regards

---

[219]With regards to the discussion in Belgium, cf. Nys (1998), 43-63; 49, and 62f. Cf. also Trouet (1998), 94f.

[220]Additional Protocol to the Convention on Human Rights and Biomedicine on the Prohibition of Cloning Human Beings. Paris 12 January 1998. European Treaty Series 168. By June 2003, the protocol had been signed by 29 states and ratified by Cyprus, Czech Republic, Estonia, Georgia, Greece, Hungary, Lithuania, Moldova, Portugal, Romania, Slovakia, Slovenia and Spain, where it is in force.

to biotechnology in general, namely that the results of research can be stored in an abstract category of "knowledge", i.e. reason, for further use.

As in the convention, the protocol avoids a clear stance on defining human forms of being. On one hand, it rebukes cloning "of human beings", defining it as either the intention to create or the deliberate creation of genetically identical human beings, while speaking of "embryo splitting" and the birth of "genetically identical twins" on the other.

3.5. The Protocol on Transplantation

The second protocol on transplantation of organs and tissues of human origin is formed on basis of earlier recommendations made by the council as well as the assembly.[221] In the preamble, some of the main COE views on biotechnology already discussed are reiterated: medical progress, and transplantation as "an established part of the health services" in particular, "contributes to saving lives or greatly improving their quality[.]" There are also inherent problems of "ethical, psychological and socio-cultural" nature linked to the field and the possibility of "endangering human life, well being or dignity" by its misuse.

The protocol is limited to the transplantation of organs and tissues of human origin, yet it excludes reproductive or embryonic or foetal organs or tissues in addition to blood or blood derivatives.

The recipient and donor shall receive "appropriate information" (art. 5 and art. 12) and all professionals shall employ "reasonable measures to minimise the risks of transmission of any disease to the recipient" (art. 6). All trafficking is prohibited by art. 22, as is the attempt to profit economically from the human body and its parts (art. 21).

As with the beginning of human life, the protocol also avoids defining its end, regarding the certification of death as matter of law.[222] Since the protocol stresses the necessity of the living donor's consent (art. 13) and the right to withdraw it at any a time, it does not allow for removal of organs or tissues on persons not having the capacity to do so (art. 14), surpassing the stance taken in the convention with regards

---

[221]Additional Protocol to the Convention on Human Rights and Biomedicine, on Transplantation of Organs and Tissues of Human Origin. Strasbourg 24 January 2002. European Treaty Series 186. By July 2002, the protocol had been signed by ten states.

[222]Art. 16: "Organs or tissues shall not be removed from the body of a deceased person unless that person has been certified dead in accordance with the law."

to research in general. When someone is considered not able to consent, transplantation may nevertheless take place, provided a brother or sister is in life-threatening need of regenerative tissue, the donor does not object, no compatible donor is available and the authorisation of the representative of the donor has been given specifically and in writing. Thus, the principle of consent is in this situation regarded as less significant than upholding the life of one's brother or sister.

It is interesting that, once again, the COE apparently ascribes a fundamental value to life, which may set aside the otherwise paramount principle of consent. In part, it seeks to protect this principle by requiring the donor not to object, awkwardly bound by the obligation to equal someone's capacity to consent with the varying legal determinations of its existence.

4. CONCLUSION

Our survey has confirmed that biotechnology indeed is formed in the nexus of academic, economic and political interests. Simply put, these interests relate either to the freedom of research or to the protection of the human being in the continuum of its development, reflecting different values and goals at stake.

Freedom of research is regarded as pivotal for the development of new techniques in biotechnology, designed to solve the biomedical problems for which traditional techniques prove insufficient. With considerable financial costs linked to this development[223], the combination of private and public funds seems indispensable and the industry has an understandable interest in recovering some of its contribution by patents ensuring return for expenditure. The COE assembly of European parliamentarians recommends, however, that plant-, animal-, and human-derived genes, cells, tissues and organs be excluded from the patent law system, since they cannot be considered as inventions.

The concept of research employed by the COE presumes the human capacity to understand the objects of analysis and by the knowledge acquired herewith to transform e.g. the insufficiency of human or animal nature. For this reason, the notion of research is often linked to the general results it is expected to produce, e.g. the improvement of health care, or its impact on specific techniques, e.g. the removal of embryonic stem

---

[223]According to some, "a drug typically takes around $ 250 million to bring to the market, and most of them fail along the way..." Bains and Evans (2001), 256.

cells from the inner cell mass of the blastocyst.[224] Defending the freedom of research is just another way of defending the freedom of the human individual, in particular its freedom of investigation and movement, because as any other collective enterprise, the word 'research' is also a name used for grouping activities of individuals. Ideally, the freedom of research is nothing else than the freedom of the individual researcher to carry out his or her business without constraint. This act of research, i.e. the study of an object in order to comprehend its nature, is at its core an act of control, since this acquisition of knowledge enables the human agent to posit itself over against the world surrounding it, so that it may make sense, and also to satisfy its general urge to know, which may be driven by the same quest for meaning.

Selecting objects for study on basis of one's interests and expecting these to be met by the very objects, are attitudes formed by a detachment of the human subject from the objects, without which the very want to penetrate into the interior of things or beings could not arise. Research on human material, e.g. the genome, is in this sense an act of self-control after detachment, by which the subject desires to understand the structures of human life, of which it personally is an instance. This is the core of the Cartesian enterprise.

From a political point of view, biotechnology propels society towards higher standards, especially in food production and health care. It is important to remember, however, that also these standards result from a reflection on the present situation undertaken by a number of individuals. They are, thus, societal ideals extrapolated from real or notional experience, which in turn poses the question of interpretation concomitant with such an extrapolation.

It is an expression of trust in human agency to conceive biotechnology as a multi-functional instrument capable of combatting disease, elevating the public health standard, improving in the quality of life and enhancing the production of animals and crops for human consumption. This trust is also mirrored semantically in the COE documents by the tendency to use the term 'biotechnology' as a grammatical subject, depicting it as almost independent entity providing knowledge as well as means on its own.

There are clear shifts in the political interpretation of the state and potential of biotechnology, oscillating between enthusiasm and rejection. The latter response stems

---

[224]Cf. Bart Hansen/Paul Schotsmans: Stem Cell Research: A Theological Interpretation. Manuscript of forthcoming publication, for a helpful clarification on this and other forms of cloning.

from the fear that the use of biotechnology will affect the human individual in an undesirable way. This reaction is not illogical, for if the human agent is so essential to the achievement of goals for which biotechnology is designed, and, thus valuable, this agent deserves a protection reflecting its value, which still presupposes, however, that some humans will endanger others by their acts. This risk surpasses the general possibility of harming someone else by one's doing, since a human being can turn into the object of research or treatment, with an increased vulnerability at particular points of its development. As long as the other is capable of consenting, i.e. competent, this danger may be eased a little by demanding respect for the principle of 'informed consent', which assumes the human ability to grasp and communicate the results of future action as well as one's will in this regard. It is a concept used in so-called 'modern' theories of state, where the human individual is a free agent possessing the inherent right to interact freely with others simply because of his capacity to do so, i.e. the inherent ability of human reason to dispose of the material as well as the means for its own activity, granting freedom of thought and conviction, sometimes understood as human autonomy.

In the juridical system, however, the principle of consent is dependent upon the legal determination of its existence, and so is the definition of the personhood underlying it, on which a European consensus obviously cannot be found. This situation is in part almost necessary, because law is based on the principle of life, embodied in human and other forms with an inherent imprecision of continuity, rendering it quite impossible for law to define life indisputably. Hence, the borders drawn by law are expressions of the human perception of reality and the conditions it places on human co-existence, which is another way of phrasing the business of ethics. While some would defend a legal positivism, allegedly removed from any form of meta-juridical reflection and, thus, independent of external sources, this position nevertheless rests on an ethical evaluation, namely that the will of the lawmaker has to be respected, i.e. that law is made in order to be followed.

When there is a lack of ethical consensus on how to interpret the presuppositions of law given by the reality humans encounter, whether physical or meta-physical, law will reflect this disagreement and become vague. This insecurity characterises the legal provisions on biotechnology, as we have seen, for its object is indeed often life in its forms. It may also explain the desire to erect structures of protection for the human being that appears as the strong agent of biotechnology, creating and changing, as well as its vulnerable object of research and misuse.

Attempts at safeguarding this protection will remain feeble, however, as long as a global agreement on the status and definition of the instances of human life called 'embryo', 'foetus', 'person', 'individual','child' and 'the incompetent' has not been reached.

This protection is usually defended by the notions of 'human dignity' and 'human integrity' employed as a short hand for practices deemed unethical. Using such short hands is convenient, of course, but not entirely constructive when the precise ethical character of an action has to be determined. While most humans would agree that in general, one should not treat the other in an unethical way, differences erupt whenever a specific procedure is evaluated as unethical and, consequently, legally prohibited. In other words, it may be quite clear that humans deserve respect for their dignity and integrity, but not what this actually entails with regards to e.g. research on embryos *in vivo*, on humans not able to consent or on those legally defined incapable of doing so, especially when this is carried out on the assumption of possible benefits generated for larger group rather than the direct benefit of the person concerned.

At a fundamental level, our survey of COE documents has also uncovered several issues that underlie the discussion of biotechnology, upon some of which we have just touched. It is in this regard significant to recall that the COE takes its *raison d'être* from the notion of human rights. Briefly put, the attempt to regulate biotechnology is an expression of the desire to ensure the protection of human rights. This makes it in turn necessary to understand how these rights in will be affected by technology in general and by biotechnology in particular with its significant direct or indirect impact on life in its forms and stages.

Curiously, the individual human being is seldom seen as the prominent motor of technology, while it is quite common to see the notion of human rights as inherently linked to the notion of the rational being and its relation to the society of which it is a member. While referring to the freedom of research, of course, implicitly acknowledges that biotechnology is brought into being by a collective endeavour of various individual acts, the increasing tendency to cast biotechnology in an active mode of speaking reveals how it is viewed as well surpassing the level of individuality.

Moreover, the wish for regulation accepts its character as a somewhat clearly defined entity that law, likewise understood as such an independent instrument, can change. Exaggerating a little, one could perceive this relation as a tension of power between two institutions in the broadest sense of the word, with law and biotechnology struggling for defining their respective space of influence and domination, still bearing in mind that these institutions only exist due to the sometimes coordinated and sometimes arbitrary combination of individual human acts, including those of the mind. It seems, then, that the notion of biotechnology somehow depersonalises the same being which the notion of human rights seeks to reinforce. This reinforcement is, typically, sustained by the notion of reason.

In fact, the notion of reason is, perhaps, the most important in terms of establishing the framework for our field. For the COE initiatives rest on the assumption that international legal regulation is possible and also necessary for protecting human rights This understanding of international law may be deeply dependent upon a modern view of the individual human being and society, for the very concept of law in this regard could be interpreted as rooted in traditions that have sought to establish human autonomy in terms of defining the sources of law and the conceptual conditions for understanding the state as a legal subject, which I shall investigate a little in the following. Moreover, the whole point of stressing the notion of human rights is the assertion of their self-evidence, i.e. that they can be fathomed by anyone using reason in accordance with its principles. Any work of harmonisation on the international level, such as pursued by the COE, also assumes that it is actually possible to reach an agreement between human beings. Hence, there must be a level of common understanding, i.e. a sphere of shared rationality, which forms the basis for seeking arguments and instruments that can be employed in order to erect structures of inspiration and protection.

From this perspective, COE initiatives appear as hallmarks of modern contract theories, applying the classical principles of the enlightenment. Regarding technology as an almost independent mechanism of societal development that somehow can be regulated implies that society indeed is a space for such regulation. Society is, thus, viewed as the expression of some general will, i.e. the sum of individual volitions, which can manifest itself through the democratic mechanisms of regulation, supervision, and support. Ironically, the individual human being loses its individuality in this process, in that it is understood as an almost formalised participant in a complex rational interaction forming techniques, technology, law and other societal units. It seems, then, that the individual human act is de-personalised, submerged by the stratification of modern society.

In this light, the struggle for reaching compromises at the level of the COE is, perhaps, less surprising. For international regulation rests upon the respect for domestic law, which ideally through the various levels of representation expresses the will of the human individual that the international regulation also seeks to influence. In other words, international law is several layers removed from the individual, which means that the attempts at regulating individual acts of, say, research will be accordingly removed, thus complicating the whole process of regulation somewhat. This may also explain the delay in regulation quite apparent through our study, for the greater the distance between the individual agent and the regulatory body, the more difficult it is to understand the actual state of art and to envision ways of changing it.

Finally, the whole idea of regulating the acts of the individual rests upon a particular concept of society, namely the modern, in which the authorities can claim access to any part that forms an alleged nation, including the mind of individual through acts of education, self-discipline and control. In a modern society, no one just does, but has to do so for a reason. I shall return to the implications of this point when discussing the possibility of biotechnology beyond modernity.

In short, then, biotechnology is in Europe seen as an array of activities contributing to societal progress, while containing the possibility of endangering the individual human being, especially in the weaker stages of its development. The human agent is conceived as a strong subject seeking to understand and change itself and the world to which it belongs through collective endeavours that depend upon the 'collaboration' of the biological material used. The notion of freedom, rooted in human reason, is pivotal to this view with its stress on the notion of consent, and implied in the use of political ideals, e.g. public health and the concept of international law.

The progress and danger of biotechnology result both from individual human acts, e.g. the application of a particular technique in order to achieve an envisioned goal, which in their complex interplay create situations that often first retrospectively appear as a blessing or curse, in particular because human agency cannot necessarily be extended to cover the organisms needed to achieve the anticipated goals. In order to grasp the character of biotechnology more fully in line with our concluding remarks, then, we need to examine the precise nature of human agency, the freedom it presupposes, the notion of technique and technology, and the wider bearings of the notion of biotechnology.

CHAPTER 3

BIOTECHNOLOGY AND HUMAN AGENCY

In the first chapter, I have situated the concept of biotechnology in terms of time, language and origin. This situation has revealed that it is intertwined with a certain view of the agent using the techniques forming the field of biotechnology. Apparently, this agent is a human being, who in conjunction with others acts upon questions and problems having been related to the experience of nature in its present state in order to satisfy various interests of research, commerce, and societal development.

As pointed out above, the various forms of biotechnology differ from technology in general by their dependence upon the cooperation of molecular mechanisms that will continue a process instigated by the human agent in some living material. At the sheer molecular level, there is no substantial difference e.g. between proteins found in humans, animals or plants, which is one reason why the two latter can be integrated into the manufacturing of medication. This agency presupposes the capacity to comprehend the reality of e.g. genes and to change it in accordance with an anticipated ideal, turning the human agent into examiner and creator. He is a subject dealing with objects.

Therefore, it is now necessary to change modes, turning from the analysis of legal instruments to the reflection on the values and understandings marking them.

It is common to see the distinction between subject and object, i.e. between the human agent and the matters he selects for his analysis, particularly as the result of Descartes' enterprise, leading to the "strong dualism of soul and body, mind and matter, that followed upon the Cartesian revolution in philosophy, and that dualism was reinforced in the eighteenth century by Jansenist traditions with French thought. That dualism severed the intimate connection of matter and biology from the rational soul-as-form-of-the body integral to Aristotelian interpretations of that moral theory."[225] This

---

[225] Sloan (1999), 56.

is another way of saying that from the 17th century onwards, (Cartesian) dualism replaces (Aristotelian) harmony, with reason as the paramount instrument of discourse and understanding, i.e. 'modern' thinking replacing medieval.

The rational self-analysis of the individual, considered necessary to establish itself as truly independent, concurs with a similar analysis of nature. Both movements carry traits of contradictii in adiecti: as man turns himself into the object of his own study, object and subject at the same time, he also ventures to separate himself from the nature he so clearly remains part of. The human condition is in this respect the hallmark of man's belonging to nature, subject to the very same limitations set by death, suffering and the satisfaction of basic needs characterising all of fauna. Hence, attempts to overcome this condition, e.g. with the aid of biomedical techniques, may be interpreted as ways of establishing full human autonomy. This is implicitly demonstrated by the view on autonomy present in the (bio-)medical setting:

> "There must be physical integrity of some sort, even in a damaged body. There must be a basic mental, psychic independence as well. There must be social relationships and or physical conditions respecting and accepting a person's intentions for autonomous behaviour. Autonomy can only find realization in the dialogues of an interactional setting."[226]

Thus appears a subject, whose strength and value do not depend upon the actual state of its body, but upon its ability to use and express its mind. Since the notion of autonomy is rooted in reason and posited in a rational discourse of agents interacting freely, it can apparently be maintained in spite of the perils of the human condition. Consequently, autonomy seems to be the very criterion for determining personhood, if not the humanity of a being, based on the concept of reason as an almost isolated entity functioning unaffected by a damaged body. In short, the autonomy of reason establishes a strong subject, which in turn reaffirms its strength by the continuous reference to the autonomy it is said to possess. The notion of 'autonomy' serves as an instrument of fortification ad intram and ad extram. In this regard, the structure of biotechnology dialectically strengthens the 'modern' concept of agency on which it relies, vastly expanding the range of its impact.

From the survey of the European context, the human subject also emerged as a possible victim of biotechnological misuse, a position potentially worsened by the lack

---

[226]Bergsma and Thomasma (2000), 24f.

of a common European definition on the human being in its different stages. Indeed, it seems that the human being is no longer understood as an ontological entity, but rather as a sequel of phenomena given names without necessity, e.g. "human life", "embryo", "foetus" or "the unborn child", gradually excluding the extremes of human existence from the concept of the human being. The latest development on the COE level, for instance, indicate a move to see the embryo as a specific entity not identical with a human being, which also has ramifications for its legal protection in terms of research.[227] The reality of human existence is in this way not seen as an expression of its inherent nature, its 'substance' in Aristotelian terms, but as the result of our perception, shifting from an ontological to a nominalist position.

Ironically, biotechnology is at the same time increasingly understood as such an ontological reality, containing knowledge, means and the potential for progress, with the concomitant danger of harming the human subject that previously had been defined as so strong. In such an understanding, biotechnology is not depicted as a name for a variety of techniques anymore, but as an independent structure expressing itself through these. In other words, it has become an institution, in the sense this term is used by Melehy, referring "to the structure of determination of the subject (the place given to the subject), to the form in which the subject is cast such that the latter remains in a fixed set of relations and exclusions; it also refers to the process by which this place is established (the giving of the place), to the productive activity of instituting."[228]

It is, thus, characteristic for such institutions to result from a deliberate enterprise of complex interaction between individual agents, while they at the same time bind these agents to particular discourses and modes of being aligned with them. In what I call "the ontological understanding" of biotechnology, to which I return in section 4, the subject is determined by the very structure it has erected, turning it into the possible victim of its own effort. As already discussed, this structure is the product of activities expressing diverse interests, giving a particular place, namely the nexus of these interests within which biotechnology emerges as such an institution of reference and determination.

Thus, (bio-)technology is somehow detached from the individual acts called 'techniques', a structure superseding that level of individual interaction where the notions of autonomy, consent and freedom would be employed to grasp the nature of

---

[227]See the proposal for the third protocol to the convention on biomedicine on research.
[228]Melehy (1997), 8.

human agency. I need to clarify, then, what would constitute agency at the institutional level of biotechnology. This clarification will be undertaken by, firstly, distinguishing the human act, technique and technology. Secondly, the modern concepts of human agency and identity will feature in a longer reflection. Finally, these themes will frame the discussion of whether 'biotechnology' is a name for practices or for a concept with its own nature.

1. THE HUMAN ACT

"Man does not always act as man [.]"[229] This point made by Joseph de Finance sums up one tradition of viewing human acts: man acts truly as man whenever his actions are in-determined, i.e. voluntary and, thus, an expression of what makes humans human, namely rationality, and not "instinctive, thoughtless movements, mannerisms, reflex actions, what is done under the influence of psychic constraint, hypnotic suggesting or demented frenzy etc. (...) human actions are the actions which a human being performs when he acts a rational being."[230]

Implicitly, this view rests upon the classical definition of man as an 'animal rationale', with a strong emphasis on the rational element of his being.[231] If our species above all is characterised by its rationality, human acts per definition will have to comply with it in order to be called truly human. In terms of agency, this position asserts that human acts can be directly governed by reason, resulting from an antecedent reflection on how to achieve an envisioned goal, with the will transmitting reason to act. Voluntarity is in this view dependent upon rationality, presupposing the will as an act of practical reason, a relation well extrapolated by e.g. Aquinas and Kant. This is not the same as acting freely, however, for "[a]n action cannot be free without being voluntary, but the fact that it is more voluntary does not mean that it is more free.

---

[229]de Finance (1991), 35.

[230]Ibid. See also A. Gehlen: Der Mensch. Seine Natur und seine Stellung in der Welt. Aula: Wiesbaden, 1986, 57 as presented in Gutmann (2002), 215.

[231]Aristotle: Metaphysics A 980b 25-28. MacIntyre (1999), 5f, points out that Aristotle did not intend to see reason as a mark of separation between humans and other animals, but rather as the specific character of our species: "*Phronesis*, the capacity for practical rationality is a capacity that he [scil Aristotle] - and after him Aquinas - ascribed to some nonhuman animals in virtue of their foresight, as well as to human beings."

What is done under the influence of strong emotion is more voluntary but less free."[232]

While rationality thus enables humans to act freely, their actions can only be carried out if they are willed, including willing what is irrational, which in this view would make them less free. Still, to will requires knowledge of the object willed, a way in which the object appears as of possible interest and benefit to me. When acting upon strong emotion, the individual may pursue an irrational goal, but the goal itself is nevertheless somehow understood rationally, as is the agency underlying it: I know that it is *that* person *I* am infuriated with or agitated by.

At this point, the interplay between reason and emotion slowly takes shape: when feeling, I also *know* that I am feeling and often also what caused the emotion, enabling me to name and to deal with it. In this sense, my reason forms an indispensable structure of reference for all the impressions made on me, employed for immediate registration as well as for concomitant or retrospect reflection. The seat of reason, namely the brain, is in this sense also the brain for registering sensations: "It is the brain that says *I*, (...) not only the "I conceive" of the brain as philosophy, it is also the "I feel" of the brain as art. Sensation is no less brain than the concept."[233]

Whether motivated by (sheer) reason or (also) by emotion, the will as an act of reason is bound by the object it wills. According to Aquinas, this is the inclination to one's own good,[234] which in his theological universe is translucent for the "universal object of will, which is the Good."[235] Plainly, without an object, there would be nothing to will, while human freedom consists in the will to either will or not to will an act leading to the object. Hence, not the will is free *per se*, but reason directing it. Human acts and omissions are voluntary, i.e. expressions of the will, except in the case of ignorance, for man can neither will nor not will what man does not know. Human willing may, thus, reflect degrees of freedom, dependent upon the intensity with which

---

[232]de Finance (1991), 49f.

[233]Deleuze and Guattari (1994), 211. To the extent that these thinkers are considered 'postmodern', their appearance at this point might seem odd. In my opinion, however, the postmodern criticism of an overly rational conception of e.g. human identity seeks to avoid dualisms.

[234]Aquinas (1995), q 1 a 4: "For everyone's will is inclined to his own good; hence, to be deprived of one's own good is contrary to the will."

[235]S Th 1a 2æ, q 9 a6, in: Aquinas (1970): "God moves man's will as universal mover to the universal object of will, which is the Good. (...) All the same sometimes God moves us to will a determinate good, as when he quickens us with his grace." Thus, the inherent tendency towards goodness is further supported by acts of divine grace.

the will is emotionally motivated rather than rationally. It may even lack freedom when emotion interferes almost completely with the functioning of reason as in rage.

Since emotions are an inherent part of the human subject, "[w]e need to be on our guard against the dualism which *separates* what is rational from what is sensible. We have not a *double* consciousness but *one only* - which is both rational and sensitive - because the human soul is *one* and the human subject is a *unity*."[236] Human freedom is not to ignore or to repress emotions, but to decide how their expression may affect the will and its acts. While an act of reason, the will is also influenced by these emotions, and especially desires, from which it cannot be separated.[237]

According to the position presented here, then, human freedom consists of the freedom that reason gives, allowing the human individual to evaluate its own emotions and thoughts and the objects it encounters. In its mind, the individual applies general modes of thinking, including the use of words and concepts common to mankind or a particular language group, to its own thought and reflection, at the same time thinking and being thought about, using no instruments for its analysis other than those provided by the very same reason with(in) which its thinking takes place. It is, in other words, thought that makes man, i.e. the *cogito*.[238] This is the Cartesian position.

Paradoxically, this individual use of the general means provided by reason can now be seen as the act truly manifesting the freedom of the individual and the value it possesses. Accordingly, the value of the individual is based on its participation in the rational condition common for mankind. This is the Kantian contribution.

If the humanity of a being relies on its rationality, acting irrationally means that man no longer acts as man. Referring to human autonomy is, thus, another way of stating that freedom and the right to its exercise. Now we can distinguish the following types of human acts:

- Human acts on a vegetative level, e.g. sleeping.
- Human acts that are determined, e.g. reflexes.
- Human acts that are voluntary, and thus (also) rational, but not (entirely) free, e.g. passion.

---

[236]de Finance (1991), 50.

[237]Ibid.

[238]The cogito itself contains the proof of its own thesis, of course. Therefore *ergo sum* is redundant.

- Human acts that are rational, hence free, and, thus, voluntary, e.g. my writing this book.

For de Finance, the two first categories would be acts of a human being, but not human acts, because they are acts we have in common with other living beings. It is difficult, however, to draw an exact line between acts on the vegetative level and those that are determined, but it is clear that they are not fully identical, either. While humans usually cannot command a reflex, which is a determined response to an external influence, they do have a certain freedom with regards to the vegetative acts inherently instigated by their nature: to a certain degree, one can decide how and when one would like to breathe or sleep, for instance. At the same time, these vegetative acts are determined, because they eventually are inevitable for one's survival. Thus, the question of determination surpasses the realm of mere automatic responses, of course, reaching into the fields of e.g. psychology and biology. Depending on one's underlying anthropology, the two latter categories may be regarded as acts of a subject, more precisely of a human individual living varying degrees of its own freedom in the ways it wills.

This argument is a petitio principii in disguise, because it uses an axiomatic definition of human life, in which humanity basically is equalled with rationality. Our review of the COE documents showed that at least one other venue is open in this regard, namely understanding humanity primarily in terms of its life, even though this venture has not resulted in a clear definition of the human being, either.

As far as biotechnology is concerned, it seems that only acts of the last category are envisioned, for the idea that research requires freedom and any procedure involving another human being its consent, rests upon the assertion of human freedom. Simply speaking, biotechnology is the common name for particular forms of human acts, ideally carried out in freedom by agents deciding to act after assessing the present and the ways of bringing about the future. Take e.g. the development of new drugs: "A wide range of discovery techniques can identify a molecular target, although genomics-based discovery is currently considered the most powerful. This 'target' is a molecular entity whose activity is considered important in a disease. The rest of the discovery process then searches for a small molecule compound that will interfere with the effects of that target. The result is a candidate drug, which is developed into the active ingredient of a medicine."[239] The human agent envisioned here displays several features:

- he defines a disease
- he understands its structure
- he employs techniques of discovery
- he isolates a particular element, e.g. the molecule
- he pursues ways of limiting the effect of that element
- he finds solutions
- he produces a drug

The notion of agency underlying this view of the human act is that of a strong subject, capable of analysing the state of others, qualifying it (the disease), using techniques rather than mere acts, discovering the world at the molecular level, forming ideals of future conditions for human existence, finding and producing solutions. In short, it possesses knowledge, insight, willpower, the abillity to execute and to cherish the fruits of its endeavour. This view of the human subject, in particular the strength of its rational power and the freedom of movement, would by some be called 'modern', since it can be easily correlated to e.g. Kant's system.

In his analysis of the primacy of freedom in Kant's work, here especially the Groundwork, G. Heath puts it crisply: "Reason as it were gives us, as rational beings, the power to stand at one remove from reality and consider it conceptually and construct ideas with respect to it."[240] The development of drugs is in this regard dependent on reason directing acts of analysis, consideration and construction.

At the same time, there is also an acknowledgment of failure and difficulties that may occur. In some areas, however, this freedom is allegedly no longer present, since structures and forces now determine human activity: the COE seeks harmonisation of regulation and enforcement, EU framework programmes are to restore European competitiveness, companies feel pressed to move into certain markets, calling for public funds, researchers are driven by the desire to see that something can be done and patients uncomfortably sense their dependence. Paradoxically, then, biotechnology emerges as based on the declaration of human freedom, but deeply marked by the

---

[239]Bains and Evans (2001), 256.
[240]Heath (2001), 53f.

sensations of its absence. This is, perhaps, the very paradox of the modern subject and its identity.

2. TECHNIQUE AND TECHNOLOGY

For an act to become a technique, a certain perspective is needed: the act has to be regarded as significant enough to study, to understand and to be repeated in order to achieve a perceived goal. If this goal is of interest to others, they will copy this act, perhaps learning it from a master as in the traditional setting of apprenticeship. The main idea operational here is that a specific way of doing will be an efficient means of per- or transforming something, as it is put e.g. in the OED: "manner of artistic execution or performance in relation to formal or practical details (as distinct from general effect, expression, sentiment, etc.); the mechanical or formal part of an art, especially any of the fine arts; also skill or ability in this department of one's art." Apparently, technique differs from a mere act, then, by its conformity, its precise scope within a certain category of human activity, such as art, and the distinctiveness of its execution.

Matters are a little more complicated than that, though. When speaking of craft, scope and skills are almost self-evident: the making of shoes, garments, utensils, buildings, furniture, vehicles, machines, roads etc. is carried out on the assumption that production actually will bring forth forms with function. Put in classical terms, technique shapes the accidentiae of the substance, which is the whole idea of design: a chair is not merely a realisation of the idea of a chair, but the manifestation of this idea in a particular instance, the skillfulness of which may increase attraction for those appreciating its quality.

Now, the designer chair may be attractive because it is either a very useful piece of work or of some aesthetic or social value, even if it may be more difficult to sit in than in a less acclaimed. In terms of fashion in clothes, this may be even more evident: *haute couture* is not necessarily very practical, since its very concept exceeds the mere protection of the body against the climate, instead celebrating the creative expression of one's individuality. Techniques may be used for simpler way of doing things, which may lower costs and thus be options for those either not able or not willing to pay the price of the proper work. Thus, it seems that for each technique, it may be possible to conceive its ideal and its defective alternatives, the latter sometimes intended to mock the first whenever something is made to look what it is not. Hence, technique is the idealisation of the act needed to achieve a goal in a way perceived as more efficient than others. This definition comprises those acts we have not been dealing with yet, namely the acts of creativity, lust or personal hygiene. Also in these cases, goals are pursued, but they are less objectifiable. It is, therefore, difficult to pinpoint exactly what makes a dish, a picture, a song, sex or a hot bath pleasant and maybe good.

Still, we mostly comprehend pleasure and goodness when encountering it directly or through its privations. For example, a bad meal usually lacks what could have been done better, which we sometimes sense instinctively or compare with what we have learned from previous experience.[241]

Finally, there are techniques of a very personal kind, the perhaps ideosyncratic ways in which I blow my nose, squeeze a tube of toothpaste, bind my laces or clean my anus, developed without much thinking, but still learned and then simply used because they seem to fit the purpose. They differ from other techniques in their limitation of scope and skills and, most importantly, in their reduced social significance as opposed to e.g. art and work, but similar to a hobby.

Apart from this latter category, we can conclude, then, that the concept of 'technique' is rooted in the ideal of an act. Although used by an individual, technique is socially framed in its dependence upon the experience of others, their willingness to share their knowledge and the common insight into the appropriateness of the ideal. In fact, the whole concept of technique is to depersonalise an act in such a way that the goal will be achieved regardless of my actual mood, subjecting myself to the technique designed to bring me further than I would be otherwise.

The relationship between a person and a technique is thus highly complex: I use a technique in order to more efficiently or appropriately achieve what I have realised to be desirable, and thus subject the technique to my mastery; in turn, I am subjected to the requirements of the technique in question. The relationship between the individual and technique thus reflects the co-dependence of master and slave.[242] At this

---

[241] These reactions also seem typical in terms of music, where people usually immediately notice singing or playing out of tune, while appreciation of e.g. transposition may be an acquired taste.

[242] It is now almost commonplace to understand this relation between superior and inferior in terms of dialectical power structures, where true control may be exercised by the apparently weak or passive part. In terms of proper slavery, the power does lie with the master, but only as long as the slave remains in its position. Once it revolts or simply quits, the dependence of the master usually becomes quite obvious. Somewhat related, a pristine moment of revelation appeared in conjunction with the final speech of president Ceausescu, where the unexpected protest of the audience made the facades of

point, the difference between act and technique may have become sufficiently clear: while the act is usually seen as an expression of the person's free will, if not indeed his freedom, technique promises the freedom of mastery by the token of constraint.

According to the OED, the term 'technology' may name "the scientific study and use of applied sciences as well as their practical application in industry." Thus, technology is here perceived as a conglomerate of ideas, knowledge and techniques employed in a specific field of science. Accordingly, it may be formed as a science of sciences as well as an application of it, i.e. as the theory and the practice of science respectively. It differs from mere technique by this wider scope. While the term 'technique' refers to a certain mode of action almost understood independently of the agent, 'technology' comprises various techniques and their specific context formed by general reflexion, analysis and experiment. For example, a Cro-Magnon specimen may have used certain techniques in carving flint-stones, but first technology will provide the means to analyse e.g. the stone quality and the techniques used to carve it. On the other hand, it is evident that any analysis relies on a preceding identification and, thus, the very existence of an object considered worth studying. One must have moved quite far away from *using* flint-stones in order to find them interesting to *study*.

It is quite unlikely that technology would have evolved without the eruption of techniques calling for categorisation, comprehension and refinement, as it did in the course of the development called the industrial revolution. With the passing of traditional industries, technology has even given name to the subsequent phase of Western development, the 'technological revolution,' bearing witness to the sustainability of this concept.[243]

Without techniques, technology would be pure science (or fiction)[244] without application, a seemingly anachronistic superstructure of speculative thought envisioning the means not *yet* available, as e.g. so intriguingly displayed in Jules

---

assumed power crumble.

[243]Here it is important to stress that both societal forms rely on technology and that both, as Jacques Ellul remarks, are based on mass production. Cf. Ellul (1990) 2, footnote 3.

[244]At this point, the question of reality emerges vividly. Pure science could be regarded as fiction, since it is about the realm of human thought, which, of course, is real, but instable. I shall return to this point later.

Verne's writings.[245] Such visions originate in the standard of contemporary techniques, just as any human idea or dream combine and project present information, knowledge or insight. In this way, technology is also the product of its own field of application, namely that of a variety of techniques, the new design of which it in turn instigates. Basically, any technique is the act of an individual, however, and technology nothing more than the sum of human activity in response to stimuli of human origin, e.g. public research programmes and the publication of new results, or of non-human, e.g. a certain molecular structure or crop transmission.

Furthermore, technology is formed by the interplay of science and technique, with science in the OED (1989) defined as "the study of the structure and behaviour of the physical and natural world and society, esp. through observation and experiment." Hence, technology is not a neutral entity apart from society and its movements, but its very product.

On this account, Feenberg is correct in criticising Heidegger's view of technology as an instrument of power almost existing apart from society, a definition reflecting the capitalist environment in which modern technology first developed. As an instrument of control, technology demonstrates its cultural embeddedness precisely in having this scope.[246] More precisely, it "embodies the fruits of normative consensus in the aesthetic, ethical, and cultural domains and not merely pure efficiency or a consumerist delirium of acquisitiveness. To fail to see this is to accept positivistic claims at face value and to exaggerate the difference between pre-modern and modern societies."[247]

Differently put, just as biotechnology, technology in general is formed in the nexus of interests originating in the objects that particular groups of human, sometimes

---

[245]For an analysis of the man/machine interface in Verne's novels, cf. Evans (1988). Cf. also the deep-rooted public fascination with future societies as presented in Science Fiction films, such as e.g the successful 'Star Wars' sequels, displaying techniques well beyond present possibilities. There are instances, where Science Fiction is more about the present situation than the future, e.g. the Star Trek tv-series of the 1960s, which with contemporary eyes are very typical products of their time.

[246] "Heidegger (...) reifies modern technology as something separate from society, as an inherently contextless force aiming at pure power. If this is the "essence" of technology, reform would be merely extrinsic. But at this point Heidegger's position converges with the very Prometheanism he rejects. Both depend on the narrow definition of technology that, at least since Bacon and Descartes, has emphasized its destiny to control the world to the exclusion of its equally contextual embeddedness." Feenberg (1995b), 16.

[247]Feenberg (1995b), 13.

accidentally,[248] will strive for. Using technology as an instrument of controlling nature has been identified as one of the key features of the capitalist system, in particular by Max Weber. This thesis has been modified by Herbert Marcuse, pointing out that more importantly, this system seeks to control labour by management.

In his examination of the relation between these two positions, Feenberg concludes that workers in the capitalist wage system "have no immediate interest in output in this system, unlike earlier forms of farm and craft labor, since their wage is not essentially linked to the income of the firm. Control of human beings becomes all-important in this context. Through mechanization, some of the control functions are eventually transferred from human overseers and parcelized work practices to machines. Machine design is thus socially relative in a way that Weber never recognized, and the "technological rationality" it embodies is not universal but particular to capitalism"[249]

In other words, in order to stimulate the work of those producing an outcome from which they do not benefit directly, mechanisms of managements are needed. Machines would in this regard also serve as instruments of control rather than merely of production, of which the now quite rare assembly line is a pristine example. This view is linked to a specific way of seeing modern society, namely in terms of its industrial components, which may be less applicable to societies now seen as mostly service-oriented[250] with less clearly manifest, but nevertheless prevailing antagonies between employers and employees.[251]

Still, the underlying point can be maintained, namely that also in societies where biotechnology is the successor or substitute of industrialisation, technology is not merely about techniques, but also about the construction of social structures that express interests and ways of having them fulfilled, e.g. by the exercise of control,

---

[248]This conjunction echoes e.g. John Locke's metaphor of the "invisible hand", according to which the acts of individuals coincide with others, sometimes producing beneficial outcomes that had not been intended or envisioned.

[249]Feenberg (1995b), 11.

[250]It could be argued that control of workforce still is a major task of management. Strategies have changed, replacing mechanisms of power and fear with more subtle forms of conviction. The focus on hard and soft skills is a way of maximising work output by refined programmes of motivation.

[251]Contemporary industries rely more than before on highly skilled staff working independently. Therefore, cooperation is a central notion in order to ease the flow of production and service. Moreover, instruments like bonus shares tie employees more closely to the success of the company employing them.

which goes for employers as well as employees. Accordingly, the philosophy of technology is marked also by societal tensions, an arena of ideological struggles, "in which technology is seen only from the culture-philosophical viewpoint of a dichotomy between scepticism vs. euphoria."[252]

We encountered the same dichotomy in the COE documents, a piece of phenomenological evidence in further support of regarding technology as formed by projections of societal ideas. Expanding on Grunwald's terminology[253], the concept of technology could simply be seen as an instrument of reflection on human agency. Accordingly, biotechnology is a distinct way in which human agency is pictured in terms of its environmental impact and operationability, regardless of its focus on human or non-human material. Therefore, it is relevant to explore the modern concepts of this agency and the human identity they presuppose.

3.3. THE MODERN CONCEPTS OF AGENCY AND IDENTITY

Biotechnology is basically about the modification of organisms or material, i.e. about change. In the course of our examination, I have demonstrated how it is created in the nexus of various interests and envisioned as a structure of human mastery, turning practices into an institution. At the same time, the techniques thus employed rely on the biochemical mechanisms inherently operational in the objects used.

As pointed out earlier, human agency is in this field not possible without such mechanisms that a little strongly could be qualified as the 'cooperation' of objects. Hence, a human organ is e.g. transplanted when it is at least probable that the body of the recipient will incorporate it. In daily life, detergents now remove protein stains better after an alkaline protease, subtilisin, has been used to produce enzymes that respond to shifts in laundry style, with coloured clothes having to be cleaned at lower temperatures. In order to achieve this, the catalytic activity and specific three-dimensional structure of subtilisin had to be known first.[254] The engineering of genes is often based on the alleged central dogma of biology, according to which the genes are encoded in DNA, copied, i.e. replicated, hereafter in parts transcribed into RNA

---

[252]Grunwald (2002), 180.

[253]Cf. ibid., 189. He only speaks of technology as an instrument of reflection, but he does link with the question of regulating actions.

[254]Harwood and Wipat: (2001), 87f.

and then translated into the amino acid sequences of the protein, mirroring the nucleotide sequence of the original gene.[255] These cases demonstrate human efforts to understand the mechanisms inherent in the structure of nature, resulting in the formulation of laws of nature, e.g. the dogma of biology, which implicitly acknowledges that there is activity independent of human doing. Various examinations of genotypes, i.e. the genetic constitution of an organism, have thus lead to the description of particular genetic codes, forming the basis for the claim that "[t]he genetic code is said to be universal since the same code, with few exceptions, is used by all organisms studied."[256] Human acts would, then, merely be "epiphenomena" of "how the genetic cards are shuffled."

Scott Gilbert identifies this human being of behavioural genetics and sociobiology as "the *Homo economicus* of contemporary economic theory. Each animal invests its genes in the next generation according to the strategies provided in the open marketplace. The hidden hand is nothing other than natural selection."[257] Such laws and dogmas imply a predictability of nature, enabling humans to maintain a mastery that otherwise might seem limited due to the immanent motion of a world that could continue without them. Human activities in biotechnology are, therefore, dependent upon insights into the way in which organisms and matter react to human intervention, leaving ample room for the mistakes caused by contingent human reasoning.

Moreover, the dogma of biology has been challenged, and especially its program view of DNA just described, because "there is not always an unequivocal relation between a given DNA sequence and a function in the developmental system."[258]

For Christoph Rehmann-Sutter, empirical findings e.g. of regulatory genes expressed in different cells without "switching on a developmental cascade"[259], do not *necessarily* warrant a reversal of current DNA concepts, but they stress that our view on DNA and its developmental function for organism is chosen rather than given, a discourse in need of genetic hermeneutics.[260] Since DNA is only one resource of development, i.e. only one necessary cause, he sees DNA as one part of a

---

[255]Cf. e.g. Helland (2000), 13.

[256]Ibid.,14.

[257]Gilbert (2002), 137.

[258]Rehmann-Sutter (2002), 37.

[259]Ibid., 38.

[260]Ibid., 43.

morphologically structured system containing a variety of elements and processes.

In terms of the human being, the body is the place where the genetic information exists, a body that is actively constituting itself and continuously developing its structural integrity, "an autonomous developmental process, which is not separate from psychic and mental and social development."[261] The inherent difficulty related to genetic data is, however, that while knowledge of it can reveal which human features may be changed, "[e]ven the changeable features involve genes that constitute the capacity for change."[262] Genes are not that stable and identity is not formed by genetic information forming a hidden essence that awaits its unfolding, but by the actual embodiment of the individual, i.e. the integral person. As Susan Oyama puts it: "we develop as wholes, not by sprouting acquired bits from a prepackaged innate core."[263] The reductionist position is further weakened when the structure of the brain is taken into consideration: roughly 100,000 genes should be able to steer ca. "100-1,000 trillion connections (synapses) between more than a trillion nerve cells in our brains (...) at least 1 *billion* synapses per gene, even if *every gene in the genome* contributed to creating a synapse."[264] This look at the neural network indicates, then, that environment and culture play very important roles in forming the trains of thought that we see as the utmost expression of our identity.

Hence, genetic data do not signify reality or probability existing undercover, they merely indicate developmental possibilities. To my mind, this state of ultimate insecurity is the very condition of human existence. In this regard, the intermittent struggle with bacteria and virus further indicates that the human ability to shape the world it encounters at the micro-organic level is far less pronounced than otherwise. In spite of this limitation, the analysis of the structures of nature, and in particular the human body, simply in order to know and to correct, represents a central feature of 'modernity'.

The collection of information about the interior of matter and beings is in this regard an act of empowerment, I think, by which the world appears as manageable rather than mysterious. For if the genes indeed were to determine human existence, its unpredictability could be overcome and the puzzle of human freedom had finally been

---

[261] Ibid., 46.

[262] Ibid., 48.

[263] Oyama (2002), 169.

[264] Ehrlich (2000), 124.

solved.

Convincingly, Allen Buchanan et al. identify a mistaken assumption at work here, namely that genetic causation will establish the truth of 'incompatibilist determinism' - "the thesis that everything that happens has a cause and that universal causation excludes freedom. Incompatibilist Determinism is not subject to empirical confirmation, however. No increase in knowledge of causation can do more than establish the first part of the (...) thesis - that everything has a cause."[265] While it might seem attractive, then, to seek to defend this thesis with genetic data, these merely serve as a fortification of the basic assumption. Genetic hermeneutics, as proposed by Rehmann-Sutter, can at this point reveal the normative presuppositions guiding the analysis of genetic data, in particular with regards to its impact on human development. Genetic observation might, thus, be highly influenced by the wish to locate the causes of the effects one encounters, hoping that a clear causality exists.

Now, the problem is that the increase in this knowledge might "lead many people to *believe* that human freedom is an illusion. In other words, it may become harder for us to *think* [my italics, LR] of ourselves as free."[266] Such a mechanism, which assumes some form of causality, is typically linked with the notion of institutions referred to above, in which agents are bound by discourses reinforcing the institutional structure.

The language of genetics used in contemporary biotechnology thus reflects a preceding normativity that, dialectically, also shapes the perception of the individual exposed to it. In this, we encounter a transformation of perception, which Jean-Pierre Wils interprets as inherently linked to the course of technical evolution: "Our habits of perception *select* phenomena on the basis of their *technological design* and *manipulate* our experiences *of* the phenomena. (...) When a mathematical and chemical formula is substituted for the body, the sensory exterior of the subjectivity of a human being, the body takes on a *phantom status* that banishes it from the world of familiar sensations and *moral attitudes*."[267] Thus, the body acquires meaning through our perception e.g. of the elements shaping it.

In their investigation of the discursive formation of the body, David M. Levin and George F. Solomon[268] identify a variety of distinct interpretations within the history of

---

[265]Buchanan, Brock, Daniels, Wikler (2000), 91.

[266]Ibid.

[267]Wils (2000), 106.

[268]Levin and Solomon (1990), 525.

medical research, which they divide into the periods of the classical age[269], early and late modernity[270], and postmodernity[271]. At its core, their thesis states a "progression in research whereby medical knowledge (a) penetrates ever more deeply into the interiority of the body, and (b) articulates the body with increasing analytic differentiation."[272] This progression is perceived as follows: in the classical age, the body was seen as an idealised projection of speculative reason, i.e. an intelligible form, combining various qualities in an organic and subtle whole reflecting the cosmological order. Medical examination was therefore prone to discover overlaps between pre-established classifications and actual observation. In contrast, early modern medicine probed into the body from a detached point of view, which allowed opening and dissecting it, e.g. in anatomical pathology. As a result, the "once sacred body, surrounded by cultural taboos, suddenly became a worldly machine, a matter of interiority, a profane flesh to been seen into and seen through, a presence conceived as if its mechanisms would eventually be transparent for technological knowledge."[273] To a great extent, new discoveries in science or technology would deliver concepts used for describing the body and its psyche[274], e.g. steam energy [275], magnetism and electricity[276], the telephone[277] and the computer[278].

With this mechanical understanding, it seemed possible to find causes for all

---

[269]Interestingly, the classical age as such is only defined negatively over against early modernity. It also covers the Middle ages and early Renaissance. Cf. ibid., 518.

[270]Early modernity is set to begin in the 17th century (ibid.), while the term late modernity seems to apply to the 20th century. Cf. ibid., 519.

[271]Here (1990) cautiously described as "very recent advances".

[272]Ibid., 525.

[273]Ibid., 519.

[274]"Even in psychiatry, Freud drew upon nineteenth-century physics in his model of libido as drive (Trieb) countered by the forces of repression, a 'mechanism' which could 'fail', or around which 'leaks' could occur. His was an essentially mechanical model of the psyche, a machine, albeit a dynamic one." Ibid., 523.

[275]E.g. the closed loop of the circulation of blood, the heart as a pump, the need to maintain blood flow in face of varying demands, blood volume etc. Cf. ibid., 522.

[276]"The therapeutic use of 'magnetism' (Mesmer) and galvanic electricity implied an electromagnetic model of inner body processes."Ibid., 523.

[277]"The telephone exchange became a model for seeing the central nervous system as a switchboard."Ibid.

[278]"The large digital computer can 'crunch' information more prodigiously than the brain, but it cannot do what the brain does. Even the analog computer is not up to the job." Ibid.

diseases, applying the simple principle of causal agency, according to which individual agents, e.g. a bacterium or a virus, cause a particular result. This process continued in late modernity, of course, but the deeper medicine penetrated the body, the more it became apparent that the concept of agency had to be revised. For instance, infectious diseases are now understood in the context of larger social and political cultures, using the language of "host environments, communicative systems, interactive fields, local economies, and planetary ecologies."[279] In the course of the same process, it also became clear, however, that "the mere presence in the body of biological agents that can cause disease (pathogens) does not inevitably result in clinical illness."[280]

Since humans respond very differently to such pathogens and in general host a vast number of microorganisms with varying capacities of instigating disease, clinical symptoms are rather an exception than the rule. With the increasing ability to predict certain genetic diseases long before symptoms may appear, Juan Manuel Torres rightly suggests, then, that the traditional way of seeing the categories of 'health' and 'illness' as complementary is quite unsatisfactory. While in the organic dimension, "to be unhealthy implies to have an illness, this is not the case in the genetic dimension. To be genetically unhealthy can never per se involve having a disease because the notion of disease and illness has been exclusively reserved for organic processes in both medical and everyday language."[281] In the genetic dimension, only deleterious genes leading inexorably to illness or those whose effects together with ordinary circumstances lead to illness should be considered as states of un-health. A person can therefore be healthy in the organic, but unhealthy in the genetic aspect, e.g. when a HIV-infected person has not yet developed any symptoms related to the virus.

The body of biochemical processes is, thus, much more subtle than the image of the machine suggests, it is a "dynamic, synergic body (...), a multifactoral network of causes and effects, in which effects can also become causes. This body cannot be represented as a 'substance'. It has become necessary to represent it, rather, as a system of intercommunicatively organized processes, functioning at different levels of differentiation and integration."[282] According to Levin and Solomon, the task for postmodern medicine consists of recognising the legitimacy of states of being as well

---

[279]Ibid., 520.
[280]Epstein (1995),10.
[281]Torres (2002), 48f..
[282]Levin and Solomon (1990), 530.

as these processes. In other words, while it is not necessary to abandon the mechanistic thinking, which e.g. made it possible to understand the heart as a pump, this thinking does need the counterbalance of system models not based on mechanistic principles, e.g. by integrating paradigms of open-ended systems as presented above, which at the time of the publication of Levin and Solomons findings still were considered a recent development.[283] In this perspective, with the individual body also seen as the social body, disease is the result of a variety of influences on the levels of nature and culture, and its treatment requires acknowledging this complex interaction.[284]

It seems, though, that such an understanding is not the major force behind contemporary biotechnology, still perceived as a structure within which human acts will change something in accordance with the goals anticipated. Whenever the result of the change complies with the intention of the agent instigating it, this change reflects the actualisation of a potentiality envisioned in the object. It is in this an act of control, by which the agent intends to create changes according to his wishes.

Already here, an intimate connection between the notions of change and agency can be perceived, albeit in nuce. In fact, the concept of 'change' is merely a specific term for a particular form of activity, namely the passing from one state of being to the other. As such, it is interjected with that of 'life', which per definition is about activity, with 'death' seen as the state of its total absence. As an absolute predicate, it lacks the degrees usually associated with qualities, the "variations in essence" in Kierkegaardian terms, a point that Anselm may have overlooked in his proof of the existence of God.[285]

---

[283]Levin and Solomon (1990), 525f: "...some current research seems to indicate the need for a genuinely new paradigm, formulated in terms of systems and processes that would be understood *non-mechanistically*: as Ilya Prigogine has argued, drawing on recent developments in physics, there are complex, open-ended systems the elements of which are so organized, so interrelated, that the conditions of the system are not predictable on the basis of the elements themselves."

[284]Ibid., 530. They here refer especially to psychoneuroimmunology, which, very basically put, seeks to discover the apparently co-dependent connections between the central nervous system and the immune system.

[285]Søren Kierkegaard critisises Spinoza for the same oversight, because the latter explains perfection by being: "the more perfect a thing is, the more it is." According to Kierkegaard, "[f]actual being is wholly indifferent to any and all variations in essence, and everything that exists participates without petty jealousy in being (...) Ideally, to be sure, the case is different (...) Highest ideality has this necessity and therefore it is. But this its being is identical with its essence; such being does not involve it dialectically in the determinations of factual being, since it is." Such a being cannot be proven, in other words. In terms of God, i.e. the old days in Kierkegaard's thinly veiled irony, this was expressed by

The notion of life is dependent upon the notion of activity or movement, while we usually do not see activity as such as a sufficient criterion for defining life, although certain natural religions and at least some scientists would do so.[286] This notion of activity is relevant for either identifying something unknown as belonging to the group of living beings ("is it moving?") or for assessing the viability of a being that is already placed in this group ("is it yet or still moving?"). Either way, we tend to think that our observation of activity in an object reveals whether life is present or not. In the words of Nicholas Agar: "Apparent self-movement or self-originated activity is, at least operationally, a very important component of 'life.'"[287] For something to change from one state to the other, however, some power is necessary to instigate that change, be it from within or from outside. In Aristotelian terms, there is no movement without a mover.[288] It is for this reason, perhaps, that we find it difficult to assume the presence of life when there is no visible change, a problem arising in conjunction with persistent vegetative states of humans.

When biotechnological means are used to achieve changes, these are rooted in the activities of individual humans collaborating, seeing themselves as the primary agents of change in this field. This ability to instigate change may stress the underlying ability to act, because it extends the exercise of control. Since humans identify activity as the central notion of determining their own existence in terms of life or death, their desire to manifest themselves through acts of change may be interpreted as a reaffirmation of their own existence: I make things change, therefore I am. Such a reaffirmation would

---

stating that "if God is possible, he is *eo ipso* necessary (Leibniz)." This is also the point made by Anselm. Really, the difficulty lies somewhere else, though, for the enterprise is "to lay hold of God's factual being and to introduce God's ideal essence dialectically into the sphere of factual being." Kierkegaard (1967), 51f, footnote 2.

[286] Cf. Dulbecco (1987), 449, ties life to the activity we call 'energy': "There cannot be life without energy because there cannot be life order without it. Life treads a thin line between order and chaos, related to its use of energy. Energy can be used to build strong chemical bonds between atoms. But life is based mainly on the plasticity of weak bonds that can be continuously broken and re-formed. Life exists in a world of weak bonds. In a world of strong bonds life is impossible because the bonds are essentially inbreakable. That is the realm of rocks and crystals, not of life." This quotation also demonstrates that the line between scientific and religious language can be very thin indeed.

[287] Agar (1997),160.

[288] See e.g. Aristotle, Metaphysics 1072b15. Aristotle's doctrine of the Prime Mover was later adopted by Thomas Aquinas.

only be necessary, if this agent had begun to doubt or to defend himself. It could also be the reaction of a human being now feeling isolated from all other sources of support.

At any rate, 'doubt' is not a concept unknown to 'modernity', on the contrary, it is closely related with the Cartesian project, for which it forms the starting point. The method of doubt, the turmoil of which is so vividly described by Descartes in the *Meditations*[289], instigates a reflection similar to the one previously undertaken in the *Discourse*, with the purpose of strengthening the cogito further. In its course, he rephrases the original version of the *cogito sum* slightly, concluding that "*I am, I exist*, is necessarily true whenever it is put forward by me or conceived in my mind."[290] This could be understood in terms of a syllogism, of which Descartes in his objections and replies on meditation two is well aware, since the premiss "everything which thinks is, or exists" would turn his argument into a fallacy. He seeks to escape this argumentative trap by referring to experience: "in fact he learns it from experiencing in his own case that it is impossible that he should think without existing. It is in the nature of our mind to construct general propositions on the basis of our knowledge of particular ones." Therefore, his argument on human existence is in a nutshell that it is "self-evident by a simple intuition of the mind."[291]

Fascinated by the faculties of the mind, he profoundly relies on it for defining the identity of the human being, neither using the notions of man or rational animal, but instead the notion of this "thing that thinks," i.e. "doubts, understands, affirms, denies, is willing, is unwilling, and also imagines and has sensory perceptions."[292] The body is implicitly present in this definition, delivering "sensory perceptions", and Descartes does see human nature as a combination of mind and body. The problem is, however, that the senses naturally can be deceived and, more fundamentally, that the body "by its very nature is always divisible, while the mind is utterly indivisible."[293] Lacking coherence, the body and its sensations do not grant any security, especially not in terms

---

[289]At the beginning of the second meditation, he describes this state of doubt in existential terms: "So serious are the doubts into which I have been thrown as a result of yesterday's meditation that I can neither put them out of my mind nor see any way of resolving them. It feels as if I have fallen unexpectedly into a deep whirlpool which tumbles me around so that I can neither stand on the bottom nor swim up to the top." Descartes (2000), 16.

[290]Ibid., 17.

[291]Ibid., 68.

[292]Ibid., 19.

[293]Ibid., 59.

of self-understanding.

In a Cartesian perspective, bodies cannot be "strictly perceived by the senses or the faculty of imagination but by the intellect alone, and (...) this perception derives not from their being touched or seen but from their being understood[.]"[294] Accordingly, it is the mind that plays the paramount role in the apprehension of one's embodied self. For this reason, Descartes appears in the sixth meditation as slightly uncomfortable with a holistic view on the human body after all, for "[a]lthough the whole mind *seems* [my italics, LR] to be united to the whole body, I recognize that if a foot or arm or any other part of the body is cut off, nothing has thereby been taken away from the mind."[295] Taken ad absurdum, it appears that the mind can almost exist independently of the body.

In a perhaps slightly ironic comment, John Cottingham notes that this "claim is revealed in its full starkness, and (to most philosophers nowadays) its overwhelming implausibility, when we remember that the *brain*, being a purely bodily organ, must, for Descartes, be as essential to the mind's continued functioning as foot or arm."[296] It is not entirely clear, however, whether the brain indeed can be regarded as a "*purely bodily organ.*" There is no doubt, of course, that the brain is an organ "that evolved by the same processes as did fins, lungs, hearts, red blood cells, stripes on a land snail, DDT resistance in a fruit fly, or the parachute on a dandelion seed."[297]

Rejecting what he describes as the Cartesian dualism dividing the material body and the immaterial mind, Paul Ehrlich underlines that most scientists[298] nowadays "assume" thoughts to require a brain, in particular the neocortex, a subdivision of the cortex, i.e. the outer shell of the brain. Without a brain, there can be no mind. This leads to the conclusion that thoughts are "physical events with (often) physical consequences beyond the firing of nerve cells[,]"[299] examples of which are mental stress weakening the immune system and the vasodilation required for an erection.

Yet, to my thinking, this conclusion does not seem to follow from the premiss, for

---

[294]Ibid., 22.

[295]Ibid., 59.

[296] Cottingham (2000), xxxi.

[297]Ehrlich (2000),120.

[298]This view is increasingly spreading among physicians, in particular psychiatrists, which allows for speaking of a neuro-physiological turn also in contemporary psychiatry.

[299]Ehrlich (2000),120.

the fact that an organ is needed for thought to occur does not entail that thought therefore *necessarily* is as physical as the place within which it takes place. Furthermore, Ehrlich states that scientists 'assume' the brain to be required for thoughts, a term that in its lack of certainty indicates the difficulty of fully understanding the nature of thought. Clearly, humans constantly analyse, remember, and reflect, often seamlessly moving from mere sagacity to penetration, sometimes ending in the state of proper acuteness. They keep thinking, "because the brain can never be shut down."[300] In that, they have "a continuous sense of 'self' - of a little individual sitting between our ears, and perhaps equally important, a sense of the threat of death, of the potential for that individual - our self - to cease to exist."[301] Ehrlich calls this "intense consciousness". In this regard, Jean-Pierre Changeux' neuroscientific observation is important: since there are no sensory endings in the cerebral cortex, consciousness is obtained without a conscious perception of the brain, on which it nevertheless depends in order to conceive itself. This means that in relation to consciousness, the "brain is an object, but one that commands everything else and that is responsible for both the perception of my body and the production of representations that allow it to be described. Even if I do not perceive my brain, naturally I do conceive of it - I can describe it on the basis of representation that I form in my brain."[302]

While Descartes ties consciousness to the mind, some, among them perhaps Ehrlich, and certainly N. Humphrey[303], would tie it to sensations, asserting that in the absence of direct sensations, "mental representations are accompanied by reminders of sensation."[304] The problem is, however, that these claims can only be tested subjectively, for the precise relation between sensation and thinking can only be experienced by the individual. This fact is also intriguing Descartes, who by the subjective examination of his thinking arrives at the cogito. In doing so, he focuses on the abilities of reason. This particular exploration expresses a preference for reason, which nevertheless seems inevitable, since human thinking necessarily is tied to this

---

[300] J.M Allman: Evolving Brains. Scientific American Library: New York 1999, 44, quoted in: Ehrlich (2000), 110.

[301] Ehrlich (2000), 110.

[302] Changeux and Ricoeur (2000), 52.

[303] Humphrey (1992).

[304] Ehrlich (2000),111. According to Ehrlich, Humphrey's hypothesis is "that sensations occur at the boundary between an organism and its environment and, in human beings, generally are registered in the mode of sight, touch, sound, smell, or taste."

realm. From the perspective of the mind, the impressions registered by the body are in themselves only of a very limited significance.

It is first when registering a sensation rationally that man may know what it is that he senses, which makes quite a difference: a strange feeling in one's stomach may, thus, be associated with nausea, stress or falling in love. In this way, humans are dependent upon thinking for identifying what they sense, while their thinking in turn may be influenced very much by these sensations.

For instance, in order to make sense of the spots on this paper that you now see, your reason needs to compute the information provided by your senses. It is too one-sided, however, to state that, "[a]ll human experience has a bodily dimension in the sense that all experience is realized via the body (...) the only means by which experience is possible at all."[305] Limiting human experience to the bodily dimension does not account for notional experience, as it occurs in the forms of reasoning or dreams, which takes place solely within the mind. One could argue, as Ehrlich does above by using Humphrey, that the material for such activities originally is derived from some real experience, involving the body, but this does not account for intuitive knowledge associated with concepts like truth or goodness.[306]

Furthermore, what really matters to humans is what they understand, perhaps even fathom, not just anything they feel, hear or see. To a great extent, empirically derived understandings provide significant grounds for human action, to the extent that, a little dramatically put, "humans would not survive if they had no knowledge about the consequences of their actions and if they had no disposition to distinguish between logically necessary and contingent empirical consequences of their actions."[307] Humans are, thus, left with the puzzle of their thinking.

A different[308] venue towards solving it, proposed by J. Fodor[309], N. Chomsky[310] and

---

[305]Dekkers (2001), 121.

[306]By now classic modern intuitionist are e.g. Moore (1991) and Ross (1939). The importance of the intuitive knowledge of the good is also predominant in Confucianism. Specifically, Mencius (Meng Tsue), one of the disciples of Confucius, argued that humans are born with a faculty of moral intuitions. Cf. Meng Tsue (1970). For a recent anthology reassissing its merits and problems, cf. Stratton-Lake (2002).

[307]Janich (2002),108.

[308]Ehrlich (2000), 114, identifies is as "a modern version of the idea that different physical parts of the brain control different mental 'faculties.' That view dates at least to Franz Gall (1758-1828), a famous german neuroanatomist and founder of phrenology, the practice of evaluating mental abilities

others, is to see the brain as composed of modules "more autonomous - isolated from the brain's background knowledge - than researchers had previously thought."[311] In this line of thought, Olaf Breidbach has defined the real brain as "built up by neurons which show highly stereotyped morphologies. That means that in spite of the huge amount of contacts each neuron has, these were not chosen at random but were developed in a complex, but highly regular structure." The point is that the activation process of a neuron is highly complicated to follow, even though the neuronal network forms a functional system, because "[a]ny activation of the system changes the system."[312]

According to Breidbach, the term 'system' is in this respect "defined by relativistic network of putative internal relations."[313] Hence, the acquisition of knowledge means that new information is incorporated into an already changing system, which raises questions of meaning[314], recognition[315], and association[316]. This is a different way of describing the heuristic problem of objectivity, in particular whether there can be certainty about the way the human mind observes the world and, perhaps, itself. It is

---

and character by examining the shape of the skull."

[309] Fodor (1983).

[310] Chomsky (1980).

[311] Ehrlich (2000), 115.

[312] Breidbach (2002), 81.

[313] Ibid., 89.

[314] Using Hegel's attempt "at a system to describe the world by employing a mechanisms of system internal reference[,]" Breidbach sees 'meaning' as "defined just by the structure in which a term is used in thought (...) The essential meaning of a term is nothing but this relational characteristic. The science of logic demonstrates that the meaning of meanings is the outcome of the relation of terms." Breidbach (2002), 91.

[315] Recognition is the identification of corresponding signals. The difficulty is that "correspondency (...) is not found by reference to an external observer but by purely system internal criteria. Any activation of the system is evaluated according to the rules by which the system reactions are defined." Ibid.

[316] Breidbach summarises his view on James Mill and his thoughts on the associative mind in the following way: "the first effects of an impression in the brain are not a proper re-representation of the physical world. The effects are described as the outcome of an superposition of inputs onto an internal activity mode of the system. (...) The result is a complex coactivation of signal pathways established by former impressions." Ibid., 92. Simply, this means that whatever I see is in my brain incorporated into a changing system comparing impression with previous experience. The spread of information in the system of the mind is, in other words, unpredictability, which allows for human fantasy and deja-vus.

also this problem that forms the stumbling block of the Cartesian mind.

Beyond doubt, understanding is the main concern for Descartes, which ideally includes all features of human existence and, therefore, also the body. The act of understanding requires a reliable mind capable of distancing itself from the experiences it encounters, enjoying independence from bodily perils. In this, humans can find certainty, yet the ultimate reliability is not given by reason itself, as Descartes elaborates in the sixth meditation, for there still is the slim chance that reality actually is not and the world and the reasoning about it merely a deception. Therefore, some sovereign expression of certainty is needed, which he finds in God. For the Cartesian subject, understanding ultimately relies on the link between reason and the non-deceiving supreme being given "in the ordered system of created things established by God. And by my own nature in particular I understand nothing other than the totality of things bestowed on me by God."[317]

Descartes' position is radical not so much in its emphasis on the significance of the mind, then, but in its insistence on its total independence over against the body to which it belongs. While this view may seem extreme, it is worth remembering that to some extent, it is also operational in the contemporary biomedical sector. The idea that bodily functions can be understood, diagnosed and treated as separate entities places some trust in the human mind. Indeed, the notion of autonomy is currently also sometimes understood in this sense, i.e. as the human ability to possess mental independence in spite of a however badly damaged body.[318] While the notion of autonomy is employed as the "best way to focus attention to the patient as a whole person and as an agent being in control of his or her own life," it means that the role of the body is neglected, for control is equalled with rational decision-making.

According to Ten Have, it almost seems "as if the autonomous subject is not embodied. Its body is merely the instrument through which the subject interacts with the world."[319] Consequently, the body is regarded as property and the notion of autonomy expresses the sovereign authority of the individual person, "with property rights over his or her body."[320] Moreover, 'modern' institutions, e.g. biotechnology, are constructed on the assumption that the human mind can penetrate reality and grasp its

---

[317]Descartes (2000), 56.

[318]See e.g. the quotation on autonomy taken from Rehmann-Sutter (2002), 46.

[319]ten Have (2001b), 67.

[320]Ibid.

nature, producing understanding.

Generally speaking, the human body is in this structure subjected to the researching mind that seeks to better what it gradually unveils. This attitude is a modern mode of being. When discussing the concept of modernity in chapter 1, section 1, I identified this mode as a state of anthropocentrism, resulting from a series of decisive developments that are characterised by the central notions of voyage, discovery, emancipation, separation, and domination. They are notions of a being exploring itself by physically and mentally approaching the boundaries of its existence. Later, this exploration would also be turned inwards in the field of psychology, a movement certainly instigated by the journey into the self undertaken in the arts, in particular in literature.[321] In short, these developments mark differentiations of a subject and other parts of reality.

Following Bruno Latour[322], Gernot Böhme argues that these "differentiations of nature and society, subject and object, body and soul have to be grasped as ones that are produced. However, it is to be assumed that the production of an objective nature, which Latour takes over from the moderns, cannot be the goal when it is a matter of one's own nature, one's body."[323] The objectification of nature is clearly seen as a modern product, i.e. of a strong subject, that cannot turn its own nature into an object, at least not willingly, for it would efface itself by this very act. For Böhme, the threat of the scientific-technological intervention, however, is to do precisely this. He supports the anthropocentrism of modernity in a Heideggerian way, situating the human subject at the centre, still "the standard by which and through which nature and non-nature are separated (...) We are this center certainly not as hybrids - e.g., as a composition of *res extensa* and *res cogitans*. (...) We are, I would say, this center in our *concerned self-givenness [betroffene Selbstgegenheit]*."[324]

Yet, the subject is not that strong, after all. Quite in line with the findings of our

---

[321]"What was thematized in Christianity as the conversion of the sinner into the saint who tells his story may be thematized in a modern narrative in a variety of ways that need have very little to do with the Christian experience. (...) To state the matter hyperbolically, we might say that every narrative of the self is the story of a conversion or, to put the matter the other way around, a conversion is only a conversion when it is expressed in a narrative form that establishes a separation between the self as character and the self as author." Freccero (1986),17.

[322]Latour (1993).

[323]Böhme (2002), 11.

[324]Ibid.

survey of COE documents with regards to biotechnology, Böhme also identifies the fundamental tension marking the modern subject: "On the one side, the body in its state, which can be determined objectively by natural science, and in its capacity to be technically manipulated; on the other side, the autonomous subject, which constitutes itself as the center of action and projection."[325] This reiterates a point made earlier, namely that in the making of its institutions, the human subject encounters itself as the victim of the instruments it has created so successfully. The energy of this almost tragic[326] development is the human quest for freedom from mundane and metaphysical contexts regarded as inhibitive. With freedom also understood as an individual act of liberation from external influences of any provenance, it is often defined as human 'autonomy', i.e. the self-governance of the human individual, an expression firmly established in the enlightenment, but rooted in a far longer theological tradition. This notion of autonomy connects identity with agency: since the human being is understood as inherently free because of the freedom of mind secured by reason, it considers itself inherently entitled to exercise this freedom through acts ideally considered as free as the reason governing them. In this sense, the human act excludes any element of *ontological* otherness: the *human* act *is* nothing more than it is. Therefore, the human agent can only refer to himself in order to justify his actions, obeying the laws he himself has made.

This notion of autonomy has been used by e.g. Tom L. Beauchamp and James F. Childress, who claim that is rests upon two conditions, on which in their view "virtually all theories of autonomy agree[.]"[327] As with all generalisations, such a claim should perhaps be understood mainly in a rhetorical sense. At any rate, they seem to understand liberty negatively, i.e. as independence from controlling influences and positively as the capacity for intentional action.[328] Hence, "[p]ersonal autonomy is, at a minimum, self-rule that is free from both controlling interference by others and from limitations, such as inadequate understanding, that prevent meaningful choice."[329]

---

[325]Ibid., 12.

[326]I use this term quite deliberately here, because in Greek tragedies, the inevitability of this development is fiercely proclaimed, which in short is the story of human self-destruction.

[327]Beauchamp and Childress (2001), 58.

[328]The difference between these two concepts of liberty has been well analysed by Isaiah Berlin: Two Concepts of Liberty.

[329]Beauchamp and Childress (2001), 58.

The notion of autonomy is in this sense typically employed over against the alleged authority of society and in particular that of the state, demanding that legal provisions do not hinder, but foster the expression of one's interests.[330] Historically, "[t]his principle was fundamental for the overcoming of the paternalistic model of medicine, which started to appear with increasing relevance since the decade of the 1950s."[331]

Accordingly, the notion serves as a means of protecting the self-determination of the individual in a context of competing interests between the individual members of society, e.g. within the health care sector.[332] For this reason, Brenda Almond argues, "[t]he idea of not having one's important interests determined by others is sometimes described as the principle of autonomy, and it means, in practical terms, that where risk is concerned, it is pertinent to ask: 'Who is entitled to choose to take that risk?'"[333]

Hence, the notion of autonomy is to protect the interests of the individual. As these interests in turn typically manifest themselves in the choices humans make, this notion is sometimes also referred to as the freedom of choice, which is seen as a value. This entails in the words of George Agich that "[h]uman action, in turn, can be regarded as free if the individual agent can identify with the elements from which it flows." Consequently, "freedom stresses the priority of the self as an individual, which is precisely what autonomy is designed to support."[334]

The concept of autonomy could therefore be understood as covering the concept

---

[330]The legalisation of euthanasia in the Netherlands and, most recently, in Belgium illustrate the vehemence of this development, in which society increasingly is seen as the provider of services demanded by the individual. The Dutch situation is comprehensively surveyed in: Thomasma, - Kimbrough-Kushner, Kimsma, Ciesielski-Carlucci (1998). Concerning the recent development in Belgium, cf. e.g. Schotsmans (2000) and Schotsmans and Broeckaert (1999).

[331]Patrão-Neves (2002), 60.

[332]With regards to the almost notorious example of euthanasia in the Netherlands, this conflict is expressed through the terms 'autonomy', i.e. the patients right to self-determination, and 'unbearable and hopeless suffering', i.e. criteria defined by society without which an act of euthanasia may not be carried out. Cf. de Haan (2002b), 169: "(...) while the patient's autonomous request for euthanasia is a contributing factor indeed, his condition is an enabling/disabling factor. Despite the patient's autonomous request, euthanasia is only permissible if the patient's suffering is unbearable and hopeless." See also by the same author (2002a), 60: " (...) one of the basic ideas of the Dutch regulation of euthanasia thus remains respect for human life, a respect that is due irrespective of whether the patient himself believes that his life is still valuable. On the other hand, the new law also tries to accommodate in some way the patient's right to self-determination."

[333]Almond: (2001), 34.

[334]Agich (1993), 99.

of dispositional freedom, which would allow for using it as neutral criterion in order to assess whether the value of freedom is protected. According to Govert den Hartogh, however, recent political philosophy would tend to see a difference between these two concepts, regarding autonomy as a so-called 'threshold concept', employed in order to evaluate whether a particular state of health represents a medical need that should be collectively met, i.e. financed. The role of health care would consequently be to protect the freedom of the individual by assuring its state of health.[335]

The notion of threshold refers in this context to the options that e.g. health care would add and in this sense, the concept of autonomy could allegedly function as a precise criterion. Thus, "[b]elow a certain level of dispositional freedom, the addition of an extra option does not make you autonomous, and beyond the threshold the addition of an extra option does not make you more autonomous."[336] In other words, certain health care provisions would not expand the autonomy of the consumer and may, therefore, not be considered part of the collective responsibility.

For den Hartogh, this assertion is difficult for three reasons: firstly, the threshold is vague, because it rests upon the distinction between needs and preferences. This distinction is made when assessing which health care provisions should be financed collectively, leaving ample room for grey areas that are not clarified when considering that some provisions will restore autonomy and others will fail to do so. Secondly, these grey areas do not even coincide. Den Hartogh uses the example of an amputation: "You can be fully autonomous while lacking the use of your legs, but it seems to belong to the hard core of our considered opinions that a medical intervention aimed at preventing an amputation should belong to the basic package [scil. of health care provisions]."[337] There are cases, then, where certain provisions seem indispensable even when they do not further the autonomy of the individual, which also here is seen as existing independently of the state of the body. Thirdly, the judgement of which options may influence the autonomy of the individual positively is dependent upon preceding judgements "of the relative importance of the options involved as aspects of a good life."[338] Extra options may, therefore, not be valuable in themselves, but still enhancing the autonomy of the individual. In sum, freedom is not the only value that should be

---

[335]Cf. den Hartogh (2000), 109.
[336]Ibid., 111.
[337]Ibid., 110.
[338]Ibid.

taken into account when evaluating the necessity of public health measures.

To my mind, this discussion shows that this particular use of the notion of autonomy rests upon preceding ethical assumptions. Almond arrives at a similar conclusion with regards to the risks to the individual's life or health that a given procedure may entail. Since these risks possibly interfere with the individual's interests, it is ethically relevant to consider them. This view rests upon underlying ethical assumptions, "for example (...) that everyone has a right to life and a right not to be knowingly or deliberately harmed by others[.]"[339]

Consequently, rights of this type are implicitly expressing the interests they are deemed to protect. When deciding, humans investigate the risks and potential benefits of the options at hand in comparison with their preferences and values, i.e. their interests. Therefore, we usually expect competent individuals to make decisions based on their knowledge of the matter at stake.

Dave Wendler rightly points out that the depth of this investigation is dependent upon the risks involved. Moreover, conceiving human decision making in these terms also assumes knowing or at least a state of conscious awareness, which includes the possibility of error. It is, therefore, possible to make decisions that conflict with one's preferences and values.[340] Such conflicts can in the strong sense of the word only be *known* to the individual itself, and even then, they would be so a posteriori, since the notion of mistake excludes making one knowingly. Therefore, it must be at least assumed that human choices are instigated by their interests, with the will translating interests into actions.

In sum, the notion of autonomy serves as a means of defending the free exercise of acts willed because of the interests they manifest, which is what human freedom is all about. This question of human freedom exceeds the limits of an alleged historical period, of course, and its constant reoccurrence is to my mind a further proof of the thesis that modern modes of being exist regardless of historical periodisations.

---

[339]Almond (2001), 33.

[340]Wendler (2000), 316f.: "Even competent individuals misconstrue their preferences and values and fail to fully understand the decision in question. As a result, they sometimes end up making choices that conflict with their preferences and values. (...) What individuals know (...) depends, in large part, upon how much they 'investigate'. (...) What decision we expect individuals would make if competent depends upon what they would know about the decision in question and their relevant preferences and values. And what they would know if competent depends upon how much they would investigate which, in turn, depends upon the risks and potential benefits involved."

In his essay on modernity[341], which includes a careful exposé of how the new meaning of freedom developed in modernity, Louis Dupré opines that the "idea of personal freedom independent of the state (...) also originated in antiquity." He especially considers the Stoic distinction between "the internal realm of freedom and the external world of compulsion"[342] influential in shaping modern conceptions of freedom, in particular Kant's ethics.

The Kantian approach has preserved its attraction and is, according to Tom Meulenbergs and Paul Schotsmans, e.g. used by "[c]ontemporary authors in bioethics [who] often build their arguments upon Kant's philosophy."[343] They rightly point out that the Kantian notion of autonomy is not identical with the way in which autonomy is used in, say, current health care, because "Kant characterises the autonomous person as someone who makes choices based on impersonal, general laws and not by reason of appeal to the person's own needs or desires." Consequently, the difference between a Kantian and contemporary notion of autonomy lies in the role of rationality: Kant does not accept any autonomous choice as morally compelling, but only "choices based on pure reason."[344] In other words, Meulenbergs and Schotsmans demonstrate that the Kantian notion of autonomy is positive and, thus, wrongly used when depicted as merely negative, as it is e.g. done by Beauchamp and Childress or in the health care settings mentioned above. For the human ability to become rational forms the very core of Kant's project. The rational state of the human being is in the making and not a static entity.

Consequently, human nature is subject to change: "[a]s nature, human beings are given to themselves; as social beings, as rational beings, as moral entities, they are objects of self-formation."[345] Crudely put, Kant is interested in what man can become, not what he actually is. It is, therefore, in my view not precise to see "Kant's attempt to characterise humans as a natural genus that is *determined* [my italics, LR] by rationality,"[346] since his point is rather that this rationality is a possibility we naturally

---

[341]Dupré (1993).

[342]Dupré (1993), 121.

[343]Meulenbergs and Schotsmans (2001), 283. They here refer especially to Beauchamp and Childress (2001).

[344]Meulenbergs and Schotsmans (2001), 284.

[345]Böhme (2002), 4.

[346]Gutmann (2002), 199.

possess, but quite as naturally tend to ignore. Otherwise, categorical imperatives would be superfluous.

It is nevertheless important to stress that Kant's view on rationality and its role in guiding human acts conceives human nature dualistically and seems to exclude the possibility of disagreement on rational grounds.

The liberation declared by Kant does not maintain that humans did not think before the enlightenment, but that their thinking was externally based. Kant understands human reason as autonomous since it provides the concepts as well as the instruments for the exercise of thought. He can base his ethics on this assumption that the insight in the common grounds of this autonomy enables us to follow the duty we fathom. Since everyone at least potentially is able to understand that the universality of ethical demands require him to act in such a way that the acts could be universally lived, ethics can be based on insight rather than external command, which is what the imperatives intend to stress.[347]

The contribution of Christianity lies according to Dupré in giving "the idea of personal freedom a wholly new content[,]"[348] since fallen man, bound by the limitations of his inadequate condition, is seen as dependent upon God to restore his former capacity. We find this idea e.g. in the work of Augustine, who puts it crisply: "For grace alone separates the redeemed from the lost, all having been mingled together in the one mass of perdition, arising from a common cause which leads back to their common origin."[349]

As I see it, stating this dependence supposes ways in which the support of God, i.e. grace, can or will be obtained: is it given completely freely, without the doing of man, with necessity, e.g. through certain sacraments, or conditionally, i.e. upon a human act? While these questions at first sight may seem of theological interest only, belonging to

---

[347]At this point, I could have included a longer reflection on Kant's system. I choose not to, since the Cartesian starting point and its historical and systematic presuppositions to my mind form the context for the later enlightenment. Moreover, placing one's trust in human rationality is really not confined to a particular period in time. In this sense, I use the cogito and its implications in order to illustrate a particular archetypical view of the human subject and its agency. It is in this regard interesting that, apparently, there is an inherent human tendency to picture the other in terms of one's own understanding rather than to accept otherness. As Richard Sprengers has pointed out, the notion of gift could substantially enrich this different approach.

[348]Dupré (1993), 121.

[349]Augustine: Enchiridion 25, 99, in: Augustine (1955), 398.

a sphere that some now regard as obsolete, it is important to stress that neither the enlightenment nor modern modes of being appeared suddenly. They result from long traditions of reflection, born by the ideas of *longue durée* substantially shaped by theological approaches.

To my mind, this is hardly surprising, for European history is fundamentally marked by the ways in which the tension between human autonomy and divine authority have been sought solved. The strong agent so clearly associated with 'modernity' presupposes a human individual grasping its own capacities in order to change the world. It is a turn towards human personhood, in which the individual understands itself as an inherently free agent, endowed with rationality and a will to execute what it fathoms.

This turn, which in the tradition of the *nouvelle histoire* typically is located at the transition from the 11th to the 12th century[350], would find a systematic expression in the thinking of e.g.Thomas Aquinas. He understands the human being as a rational and social animal, created by God. Unlike in contingent human life, there is in God no difference between actuality and potentiality, because this difference expresses a defect in existence. Hence, whatever he wills, happens.[351]

Since God per definition is not bound by time, neither is his will that is an identical expression of his being. This divine will is an eternal law, which is "nothing other than the exemplar of divine wisdom as directing the motions and acts of everything."[352] The ideas existing in God have the character of a 'law', a term that Aquinas defines in a twofold way as a judgement of practical reason given by a sovereign governing a whole community[353], i.e. as a governing principle, and as reason as such, i.e. as an evaluative principle. In the latter case, the notion of law presupposes the formality of reason ruling will, i.e. a formal cause.[354] God is, thus, conceived as a rational being, too.[355]

Humans can participate in the divine wisdom with their theoretical and practical reason by virtue of the natural law, *lex naturalis*, which is "nothing other than the

---

[350]Cf. the by now classical thesis of LeGoff (1964). For his understanding of history as developing under the influence of ideas and social movements, cf. LeGoff (1988).

[351]1a 2æ q 3 a 4, in: Aquin (1966).

[352]1a 2æ q 93 a 1, in: Aquin (1964).

[353]1a 2æ q 91 a 1, in: ibid.

[354]1a 2æ q 90 a 1, in: ibid.

[355]1a 2æ q 90 a 1, in: ibid.

sharing in the Eternal Law by intelligent creatures."[356] This eternal law of God is made known to humans through revelation (lex divina) and reason (lex naturalis). God proclaims the natural law by directing human reason in such a way that everyone can know the nature of this law simply by making use of the facilities given by being human.

As such, i.e. as rational beings, humans have the ability to use reason in accordance with its own precepts. When doing so, they use reason in accordance with divine will and, furthermore, they gain insight into its provisions. Human reason can, therefore, obtain natural perfection on its own account, "by which, of course, is not excluded the work of God, who acts interiorly in every nature and will."[357]

The natural law is natural, because it gives humans the ability to act in accordance with their nature, i.e. to use reason in order to determine criteria, goals and means for human agency. In order to reach supranatural goals, however, divine revelation is necessary. It is first by the theological virtues of faith, hope, and charity, given through grace alone, but requiring human acceptance, that reason can reach its supranatural perfection, then.[358] In Aquinas' thinking, this stipulation is important, because he identifies the human tendency towards transcendence as an inherent element of human existence. It would for him be difficult to assume that humans could strive for perfection without paying tribute to this side of their own nature. In this, he reveals his Aristotelian heritage, regarding man as much more than solely mind or body. Human nature is seen as to be discovered over the course of time, realising its potential when using the means it provides itself.

The word 'nature' is used in connection with the concepts of natural law and the laws of nature alike, which could cause confusions. Therefore, Aquinas defines the term 'nature' as either pertaining to something set by nature or to nature not demanding the opposite of something.[359] In the first category, Aquinas uses an ethical provision as example, namely that no one is to be hurt by others. This is naturally so, because the first principle of practical reason rests upon the meaning of the word 'good', which it seeks ways of realising. Goodness is the ultimate goal at which everything aims and the

---

[356] 1a 2æ q 91 a 2, in: ibid.

[357] 1a 2æ q 68 a 2, in: Aquin (1974).

[358] 1a 2æ q 68 a 2, in: ibid.

[359] 1a 2æ q 94 a 5, in: St. Thomas Aquin (1964).

first command of the law is, therefore, that the good be sought and evil avoided.[360] This is a self-evident imperative, according to Aquinas, reflecting the nature of reason, which is to understand the principles forming the basis for its application. Upon these first principles, of which goodness is one, theoretical reason rests axiomatically, i.e. their nature can be grasped through intuitive insight. Humans apply the concepts provided by theoretical reason when using practical reason, which is individual and contingent, but dependent upon our final goal, which is the common good. In the second category, Aquinas places e.g. nakedness, which is natural, because nature does not provide any clothes. Similarly, common property and universal freedom are provisions of natural law, since private property and slavery exist due to human endeavour and not in nature as such.[361]

Theoretical reason provides humans with the knowledge of certain general principles rooted in divine wisdom, but not necessarily with insight into every truth. Practical reason can also formulate principles, but not all individual precepts. Therefore, human reason has to continue further, using the general rules and goals for human activity contained in it. These acts are the concerns of human law, formulated by the whole people, i.e. communitas perfecta[362], or its representatives on basis of the natural law. Since human laws can only look at outward and observable attitudes, i.e. that which can be communicated, it cannot deal with the nature of things.

Additionally, given the inherently individual and contingent character of human acts, human law is fallible. It is divided into positive law (ius civile) and justice (ius gentium). Civil law is an expression of the will of a political community, which means that it is a construction of mutable and diverse character. International law, i.e. the law of peoples, rests upon the principles of the natural law. Accordingly, justice in trading is necessary, so that humans can form the social communities on which natural law depends, because humans are social animals.

The human agent depicted by Aquinas enjoys the liberty to understand the conditions for its actions as well as autonomy in the strict sense of the word, i.e. the ability to govern itself by the appropriate use of reason. Human laws are also expressions of this self-governance and it seems oddly topical that Aquinas should stress that they cannot pertain to the nature of things. In the COE documents surveyed,

---

[360] 1a 2æ q 94 a 2, in: ibid.

[361] 1a 2æ q 94 a 5, in: ibid.

[362] 1a 2æ q 90 a 3, in: ibid.

difficulties appeared whenever regulations ventured to do so, e.g. in terms of defining the human being. It seems, then, that we can ascribe certain qualities to humans and attempt to prescribe means of their protection, but not fathom in distinct legal terms what it means to be human.

Since God for Aquinas is omnipresent and sapient and non-sapient nature secondary causes of divine activity, man has in fact been liberated to trust his own efforts as long as they comply with the order of reason. While some would interpret this as the first sign of secularisation, it could be argued that the very nature of Christianity consists of liberating man in this regard.

Considering that these thoughts originate in the 13th century, it should be clear that modern modes of being have existed in other times than modernity. Indeed, one could draw a line from the Stoa over Aquinas to Kant and from there into other ways of seeing the human agent as capable of understanding and changing the objects it encounters, with the catholic magisterium remaining a stronghold of this modern approach in its insistence upon the ability of man to grasp and enact ideals such as justice and peace. Assuming that even tainted human nature can be directed by the splendour of truth, the magisterial teaching underlines the role of reason in identifying the truly human way of life.[363]

Thomistic thinking was revived e.g. in the Spanish neo-scholastic period, where it served as a means of examining the nature of law, in particular with regards to its international character. Franciso de Vitoria (1483-1546) equals *ius gentium* (i.e. international law in current usage) with the law existing between peoples, *ius inter gentes*.[364] There is a natural interdependence between humans, with mankind as the greatest form of society possessing political rights in itself due to natural law. Thus, the idea of the *orbis christianus* is replaced by a humanist approach, where reason is the source of law establishing a universal community, seeing the whole world as one republic, *aliquo mode una respublica*. Both international and constitutional law belong to the realm of ius gentium, resting upon lex naturalis, and they are therefore not

---

[363]A prominent example is John Paul II (1993), 88: "The*morality of acts* is defined by the relationship of man's freedom with the authentic good. This good is established, as the eternal law, by Divine Wisdom which orders every being towards its end: this eternal law is known both by man's natural reason (hence it is "natural law") and - in an integral and perfect way - by God's supernatural Revelation (hence it is called "divine law")."

[364]Cfr. Hadrossek (1965), 824.

subsumed under ius civile, which as positive law in the Aquinian understanding merely expresses the will of the law-maker.

In terms of the individual, the main importance lies in its ability to have an immediate insight into the character of act through its conscience, i.e. the ability of practical reason to distinguish right from wrong, rather than merely obeying the provisions of (positive) law.

In the view of Francisco Suarez (1548-1617), reason alone can judge what is right or wrong. Therefore, law qua law does not bind the individual and in particular not its conscience, for there can be doubts about its provisions and *lex dubia non obligat*. Law cannot claim possession of one's conscience, because it is 'owned' by one's freedom. In cases of conflict, the inherent human ability to choose and act, rooted in in-determined human freedom, is superseding. Therefore, law and conscience are opposed. A single doubt about the appropriateness of a law would thus warrant not respecting it.[365]

Gabriel Vazquez (1549-1604) would not go that far, but law is for him a rational act, too, and therefore not a pure act of will. It is an intellectual judgement, dependent upon the natural law preceding any act of reason or will, even with God. In fact, lex naturalis is in this sense identified with rational nature, i.e. reason in itself, with which individual acts can correspond or differ. God's eternal law is, therefore, not really law, but an exemplary idea after which God creates rational nature and its human embodiment.[366]

These three thinkers demonstrate how the originally theological enterprise of understanding the will of God gradually changes its focus to the abilities of the human subject, in particular its rational capacity. In this regard, the secularisation of the human mind and its societal contexts of the *anciens régimes* was effect and cause at the same time, dialectically reinforcing what it had been stirred by, namely the declaration of human autonomy over against any authority other than that of reason. If there is divine light in humans, there is no need for them to go anywhere else in order to find justification for their acts. Thus, in the course of the enlightenment, it was a mere step to declare that reason can substitute God, with science propelling this process. With the reign of reason, it was believed, facts would erase doubts and mere beliefs.

In his analysis of the enlightenment, Peter Gay characterises this development as

---

[365]Mahoney (1989), 227f.
[366]This presentation relies on: J. Hellin (1950), 2609f.

the "(...) irresistible propulsion of modern scientific inquiry (...) toward positivism, toward the elimination of metaphysics, and the clean separation of facts and values, foreshadowed by Bacon, implied by Newton, triumphantly announced by Hume, taken for granted by the leading scientists of the late eighteenth century. Scientific thinking exacted the stripping away of theological, metaphysical, aesthetic, and ethical admixtures that had been a constituent part of science since the Greeks (...)"[367]

The theological debate just sketched provided a setting also for this part of the enlightenment. In Dupré's opinion, a significant impact in this regard comes from Ockham and his theory of the absolute freedom of God. Freedom in its absolute sense is neither bound by any necessity, nor by rationality. It is, indeed, sheer will, i.e. voluntarism: "Divine power remains absolute at any moment, supporting the ordained - that is, the actually chosen order - yet exceeding by its infinite possibility the scope of that actuality."[368] While the Aristotelian position operated with a divine model of rationality, also to be found in Aquinas, "the voluntarist conception was modeled after the archetype of an open-ended, indefinite, and hence potentially infinite power."[369] Consequently, law could be based merely on the will of the law-maker, and with the absence of a predictable order in nature or reason, "[t]ruth was not simply given with insight; it required an assent of the will. The knower assumed full responsibility for what he or she affirmed."[370]

Dupré's observations are interesting, I think, because they contextualise the discussion of human autonomy by revealing its fundamental tension between understanding reason as bound by its own theoretical necessities on one hand, e.g. resulting in the formulation of the *lex naturalis*, the *contrat social* and categorical imperatives, and the will displaying its utter arbitrariness on the other, epitomised in the assertion of full human autonomy. In this sense, our discussion could be rephrased as an inquiry into the freedom of the will.

The idea that the will at least potentially possesses infinite power has had some effect on the development of biotechnology. In chapter 1, it became clear that the human agent is regarded as at least partially able to change what he has acquired knowledge of, e.g. with regards to the genome, and that this potentially harmful power

---

[367]Gay (1996), 159f.
[368]Dupré (1993), 123.
[369]Ibid., 124.
[370]Ibid.

needs regulation in order to protect the weak in society. Also, our reflection in this chapter has shown that mental independence is assumed in spite of the physical state of one's body. The freedom of mind, secured by reason, is in this respect the condition for asserting human freedom, and the acts that it governs through the will are regarded as free, too. The role of the will is, therefore, to instigate the actions that reason envisions. It is an act of practical reason insofar as humans can only *will* what they comprehend, which makes the will the executor of reason, i.e its expression rather than an element of it, although like reason in general, it may be stimulated by emotions.

Descartes understands the inherent connection between reason and will like this: "As for the faculties of willing, of understanding, of sensory perception and so on, these cannot be termed *parts* [my italics, LR] of the mind, since it is one and the same mind that wills, and understands and has sensory perceptions."[371] This would also make the will quite as free as the mind it expresses, a position that seems to have been contested quite as much as defended, the latter typically by those analysing the human condition on modern terms, e.g. Darwin, Freud, and Marx.

In a theological perspective, it is sometimes maintained that the most fundamental difference between Protestant and Catholic theology lies in this anthropological query. In our present context, it will suffice to point out that the human will of course is bound by the objects it encounters, since they present the necessary condition for making choices. While the human will in theory, then, is free to make choices, it is in practice always bound, for whenever I will, my willing is tied to the object.

In a slightly unusual argument, Ehrlich presents this old problem in the light of neurological findings:

> "That enormous complexity of our brains can also (...) explain humanity's famed 'free will.' No modern computer (...) could calculate fast enough to specify all the possible trees of interactions that can be generated by the nerve impulses that travel among the many billions of nerve cells in your brain in a few seconds. Each nerve impulse would occur with a certain probability, would have a 'cause,' and would contribute to one of a probabilistic array of effects. It is, then, only in a probabilistic sense that every move we make could be predetermined, but those ultimate 'causes' of apparently free choices can never be traced, and we can never be aware of them. Thus, although in the abstract there may be no free will, in practice the brains of human beings evolved so that intentional individuals can make real choices and can make them within a context of ethical alternatives."[372]

---

[371] Descartes (2000), 59.
[372] Ehrlich (2000),125.

The speed and complexity of nerve impulses would thus warrant that our choices at least appear free, while they may not be so, which according to Ehrlich would not really matter since we cannot be aware of any determination anyway. It is almost ironic that the mind seems incapable of understanding its own working at this level, which includes not knowing whether our thinking indeed is *determined* and not merely conditioned by the structure of our brain. Still, the mere fact that humans can reflect on this question is an albeit weak indication of a certain degree of freedom in this regard.

In current biotechnology, the notion of autonomy refers to the freedom of mind and to that of the will, stressing the right to human self-determination simply because of one's freedom. In this sense, this notion combines the traditions of natural law and voluntarism by stating that humans can reason for themselves, possessing an inherent freedom and exercising their decisions freely by virtue of a will at least ideally unrestricted. The presuppositions at work here are, then, that basically, humans are free, they therefore do what they will, and since they are *rerum novarum cupidi*, they keep exploring, with their research marked by the want to do what they can, which in turn necessitates taming this drive.

I consider the distinction between 'will' and 'want' relevant, because the want to do is not necessarily rational, which may explain its continuous vigour, while the will as such is an act of reason that can be ruled rationally or emotionally. This leaves room for unpredictability, making it obviously difficult to regulate, which is further impeded because "technological development is a collective activity. There is no specific willing person or 'general will' in charge. Due to the large number of contributors and the large (social, geographical en (sic) temporal) distances between them these contributions are not co-ordinated very well and often counteract each other or push in different directions."[373] In other words, if we had clearly defined agents carrying out acts causing particular effects, as in road traffic, it would be possible to establish rules and enforcement. As the increasing problems in traffic control demonstrate, however, even this fairly specified area cannot be regulated easily, mostly because human behaviour remains multifactorial.

The development of (bio-)technology takes place in a far less confined space with individual agents acting upon a variety of interests and stimuli in a collective interaction marked by competition and ambition. One could get the impression, then, that

---

[373]Hogenhuis and Koelega (2001), 210.

biotechnology is not really willed, but just happening. Such a view would fall short of seeing that even the complex interaction resulting in contemporary biotechnology is carried out by individuals performing sequels of depersonalised individual acts, i.e. techniques, mostly in cooperation with others.

When relating the notion of the human act to common views on biotechnology above, I concluded that the human agent is depicted as possessing knowledge, insight, willpower, the ability to execute and to cherish the fruits of its endeavour. Moreover, the very notion of technique presupposed that human acts willingly are repeated in a particular fashion and the emphasis on the freedom of research further supports this role of the will. Without individuals acting at the academic, economic and political levels, there would be no biotechnology. At the same time, asserting the need for regulation in view of possible negative effects implies that human acts can cause what they were not intended to do. Thus, voluntary and free acts form together with the notion of rationality they entail a necessary, but not sufficient condition for understanding the present shape of biotechnology. At this point, then, we need to analyse whether biotechnology is merely a name for human acts, which would make it a voluntary result, or an entity with a nature on its own.

4. IS 'BIOTECHNOLOGY' MERELY A NAME?

'Biotechnology' has appeared as a word so conveniently used to label and, thus, to categorise a complex structure of research and procedures, i.e. 'techniques', in the fields of medicine and biology. The convenience lies in perceiving a clearly defined entity that may be analysed, discussed, and administered *as such*, in the same way that the academy systematises specific questions or interests by disciplines, fields, and faculties. Biotechnology has proven to be no such clear entity, though, since it is shaped by a diversity of means and objectives in the nexus of academic, political, and economic interests.

There is one common stimulus for these diverse practices, however: biotechnology is employed with a strong belief in the human ability to better what is deemed imperfect in flora, fauna, or humanity itself. Simply put, like any other form of technology or human activity in general, *bio*technology is basically about change, more precisely about the transformation of actuality into its scientifically assumed potentiality, with the human agent as its prominent executor.

Evidently, the very depiction of an inherent potentiality is intertwined with the idea of normality, namely in that viewed from a fixed point in time, any past, present or future state of existence may be regarded as the standard to which those things or beings not yet corresponding need to be elevated.[374]

An example is the notion of public health, by which some humans distill the ideal of a lifestyle considered 'healthy' as opposed to other, albeit less viable options, from the various individual instances of human life in a given society and its time.[375] Often, the ideal human life thus envisioned is one in equilibrium, regarding illness as an instance of yet another biographical hurdle decelerating the actualisation of human potentiality. Interestingly, illness, while an inherent element of contingent existence, is understood as *contra naturam* due to its 'attack' on idealised human life, postponing the recultivation of paradise further. As in English garden architecture, nature in its imprecise diversity is perceived as flawed by itself, a misfortune in need of special treatment in order to become truly natural and, thus, pleasing again. It is first by the hand of man, then, that nature can become perfectly natural.[376]

Consequently, the actual course of nature, which might be willed by a loving God,[377] is interrupted by the human being considering itself detached from it after having objictified nature in the long process of de-mystification described by e.g. Max Weber.[378] New procedures, such as culturing embryonic stem cells, instigate hopes

---

[374]The use of the term 'normality' is at this point not entirely precise, in that it could insinuate that enhancement of e.g. crops in itself were an act of normalisation. In reality, it results from a far more complex process, however. The question at stake here is, thus, whether a fascination with the realisation of potentialities expresses discontent with actuality or arises independently from it.

[375]It is not unusual to define a (public) health policy like this: "A health policy (or public health policy) may be defined as a policy which, through the use of various policy instruments, attempts to promote the health of the population. These policy instruments may take the forms of regulations, incentives or disincentives, communication or information." Theofilatou (2000), 19.

[376] With regards to the rationale behind English garden architecture in the 19th century, cf. for the great lines of the development e.g. Enge; Schroër; Mataj; Classen (1990). For the English garden in particular, cf. Brown (1989). In the same respect, domesticising animals could signify a similar attitude towards non-human nature.

[377]From a theological point of view, it seems to me that the connection between the course of nature and the will of God has to be expressed cautiously, at least as long as the actual meaning of death and suffering has not been grasped fully. Therefore, I am not sure that an Augustinian certainty of the following kind can be applied today: "Nothing (...) happens unless the Omnipotent wills it to happen. He either allows it to happen or he actually causes it to happen." Augustine: Enchiridion 24, 95, in: Augustine (1955), 395.

[378]Cf. e.g. Weber (2002).

hitherto unknown: "Will this new medicine of 'regeneration' empower society to eliminate the vicissitudes of the natural genetic 'lottery,' with everyone able to use his or her own cultured cells as a source of new tissues and organs?"[379]

Against this background, biotechnology is a proclamation of human capacity over against the insufficiency of nature, erecting a structure conditioned by nature and its principles, yet resulting from heterogeneous initiatives undertaken to overcome these very conditions marking it, thus completing the emancipation of the now strong human subject by virtue of its own imaginative rationality. Employed to solve the problems puzzling an individual or collective mind, at times resembling play in that knowledge or practice is sought for its own sake,[380] biotechnology truly serves as a human riddle, finding right answers to own questions.[381]

The very presupposition of biotechnology, namely that humans somehow are called to change crops, animals or humans in order to optimise them in accordance with an anticipated ideal state, marks a fundamental difference to other forms of technology. For other forms of technology are about objects that do not fall under the dominion of life, but of sheer matter. While dependent upon micro-biological processes, some of which we have described in chapter 1, section 2, the human agent nevertheless seeks to direct these towards goals that by the same token manifest the desire to reinforce its position of control. The true challenge lies in demonstrating dominion in terms of the complexity of life rather than, e.g., in the advanced production of steel, after all. Thus, the penetrating analysis of human and non-human life at its micro-organic level results in a sequel of riddles that apparently need solutions not yet found. In this regard, the fervour accompanying and stimulating the biotechnological enterprise is quite understandable, for life in its richness is, indeed, a source of maze that non-biological products cannot equal.

It is important to bear in mind, however, that in spite of some public perception,

---

[379]Lenoir (2000), 1425.

[380] A point already made, albeit indirectly, by Thomas Aquinas in 1a2æ, 1, 6 of the Summa Theologiæ: "Even the disinterested knowledge of theoretic science is loved as good for the one who puruses it; moreover it is caught up in the complete and perfect good which is his ultimate end."

[381]See e.g. the opening speech of the Danish minister of economy and commerce, deputy prime minister Bent Bentsen, at the first Belgian-Danish Forum for Innovation in Biotechnology, Brussel 29th May 2002: "Biotechnology strives to answer fairly fundamental questions. It is our responsibility to show people that the aim is not to alter nature just because it is possible. Biotechnology is about using nature in the most optimal way for the sake of a better future for all." Unpublished manuscript, p.6.

biotechnology is not an entity with substance, existing as such, but merely a name for certain human practices in the field of technology that "is just another dependent social variable, albeit an increasingly important one, and not the key to the riddle of history."[382] If technology is, indeed, a variable, so is biotechnology, and its function as a a key to the riddle of nature equally limited. Moreover, as a human activity, biotechnology is marked by the usual expressions of human contingency, which is (almost) self-evident. It is, therefore, variant *and* fallible.

Having developed concomitantly with the medical and agrarian industry, the use of biotechnology reflects the general desire to interpret development as progress. Both of these notions are essential to the 'worlds of welfare capitalism,' as Western societies have been aptly classified by Gøsta Esping-Andersen: "Here it is *technology* (my italics, LR) which is the driving force. Our theoretic heritage emerged in an epoch (...) in which social and economic progress, liberty, and modernity were associated with the dismantling of the absolutist, interventionist, and authoritarian state."[383]

In the course of abolishing the *anciens régimes*, and the structures identified with them, technology appeared as the instrument and objective of the now bourgeois societies, in which technicians in the 19th century advanced to artists, perhaps best illustrated by the commonplace of marvelling major railway-stations as "new cathedrals",[384] which art in turn reflected directly by choosing technical objects as motives or indirectly by its focus on nature still unspoilt.[385] After WW I, in which biotechnology had shown a devastating potential in the chemical warfare subsequently

---

[382]Feenberg (1995b), 8

[383]Esping-Andersen (1991), 222.

[384]One of the most telling examples of this is the railway station in Cologne, which on purpose was built just next to the cathedral, the completion of which itself intended as an expression of German unification and technical enterprise. Viewing stations this way is a commonplace, to the extent that this image is no longer set in inverted commas and used widely, e.g in the presentation of a new ministerial building in Copenhagen, taken over from the Danish East Asiatic Company (EAC or ØK). Nielsen and the Danish Ministry of Food, Agriculture and Fisheries (2001), 18: "So the new cathedrals of the railway era were towering above the surrounding houses in Europe and the rest of the world where the Europeans were gaining ground." The image has also been used for e.g. industrial plants. See e.g. Ebert and Bednortz (1996).

[385]Jean Monet and Claude Manet form an interesting pair in this regard. For a very competent introduction to the impressionists in general, cf. Bretell (2000). Concerning art in industrialised societies, cf Bretell (1999). With regard to the urban aspects of this development, cf. Craske (1997).

banned, the role of technology increased further, especially in the Soviet and Fascist systems with their respective attempts to create the new human being.[386]

It is hardly surprising, then, that technology also was the paramount means of societal development in Europe after WW II, propelled by the Marshall plan aid in Western Europe[387] and the will yet again to create different societies in all its parts. This time, it should be achieved with entirely new forms of multilateral cooperation to preserve peace, e.g. within the originally Western frameworks of the Council of Europe and the Roman Treaties for Coal and Steel with its later transformations into the European Economic Communities (EEC), the European Communities and the European Union.[388] The 1958 world exposition in Brussels with its hallmark of the atomium is an exquisite expression of this belief in the inherent connection between societal progress and technology. Having been the object of a longer public debate about its future use and need for renovation, this tainted symbol had by 2002 become a museum of itself, hosting an exhibition of the art of living in the 1950s and the 1958 exhibition instead of a restaurant. At least in terms of the trust in atomic and industrial technology, times have changed and enthusiasm has clearly shifted towards our field:

> "Biotechnology is not just about recombinant DNA, of cloning and genetics: it is equally about producing more prosaic materials, like citric acid, beer, wine, bread (...) It is also about providing clean technology for a new millennium; of providing means of waste disposal, of dealing with environmental problems. It is, in short, one of the two major technologies of the twenty-first century that will sustain growth and development in countries throughout the world for several decades to come. It will continue to improve the standard of all our lives (...) No aspect of our lives will be unaffected by biotechnology.[389]

---

[386]In the Soviet system, the belief in the new human being was not shattered by the experience of WW II. In a booklet produced for the world exhibition 1958 in Brussels, the continuity between the victor of the civil war and the protagonists of contemporary science is stressed. Abteilung der UdSSR auf der allgemeinen Weltausstellung in Brüssel: Schöpfer und Erbauer des Neuen. Die wissenschaftlich-technische Intelligenz der Sowjetunion. 1958.

[387]On 3rd April 1948, President Harry Truman signed a decree that provided a financial aid of 12 billion US dollars to the European countries. This plan had been presented by the secretary of state, George Marshall, in June 1947, originally also comprising states under the influence of the Soviet Union.

[388]The public use of European terminology is not always precise. The term 'European Communities' refers to the three economic communities based on multilateral trade treaties. The term 'European Union' expresses the political character of these communities, which is a distinct, but not clearly separate issue.

[389]Ratledge and Kristiansen (2001),1.

Some even conclude that DNA is being dealt with in a quasi religious fashion, since "the contemporary rhetoric of DNA fits the categories of the medieval rhetoric of the soul. DNA is that which constitutes the essence of our being, that which determines our behavior, and that from which we can be resurrected after death (à la Jurassic Park)." One reason for this attitude is, apparently, that "[s]cience has become the rock of unchanging truth in our tumultuous time."[390]

Similar roles were anticipated for the nuclear and industrial sector in 1958, but the long European crisis of the 1970s displayed the fragility of this new world, the ruins of which form bizarre landscapes e.g. in parts of Lorraine.[391]

Still, the technological techniques have become an integrated and helpful part of daily life in Europe, spanning from IT over automated production and mining to humanlike insulin. Without technology, daily life would indeed be much more cumbersome. At the same time, there is a certain uneasiness about this development, as clearly demonstrated by the ecological movement and its political expressions.[392] In addition, it seems that previous conceptions of human identity, societal interaction and economic development can only be applied with difficulty, if at all.

---

[390] Gilbert (2002), 136. He refers to the work of Dorothy Nelkin and Susan M. Lindee: The DNA mystique. The gene as a cultural icon. W.H. Freeman: New York 1995. However, the exact reference is missing in the book, with its reference lists marked by several omissions.

[391] In this regard, even a brief look on international train connections reveals interesting features: From 1953, the "Montan-Express" 231/232 served Frankfurt-Coblence-Luxembourg, bearing witness to the emerging cooperation of coal and steel production. For the same reason, a French motorcar ran as express train 1101/1124 since autumn 1952, serving Frankfurt-the Sarre-Metz/Bar-le-Duc: "Eine weitere Verbindung erblickte am 5. Oktober 1952 das Licht der Welt, die die Keimzelle eines späteren FT-Zuges [Ferntriebwagen, i.e. long distance railcar, LR] bzw. TEE [TransEuropExpress, LR] werden sollte, und die in der Zukunft häufigen Wechseln sowohl des Weges wie der Verkehrszeiten unterworfen worden war, die eigentlich trotz der engen Verflechtungen zwischen den beiden Nachbarländern nie richtig leben, aber als hochqualifizierte Zugverbindung auch nicht sterben konnte. Gemeint ist die mit Schnelltriebwagen der SNCF eingerichtete Verbindung 1101/1124." Scharf and Ernst (1983), 184. This connection was turned into a more prestigious TEE in 1970 and still exists as a EuroCity connection, but the route has never been a great commercial success. It is also telling that of all high speed railway lines in France, the Eastern line to Metz and Strasbourg, although in the planning for some time, has not been constructed yet. A more general overview of the French post-industrial remains is given in Dama (1980).

[392] Cf. e.g Hösle (1991); VandeVeer (1994); Ferry (1995). Elliot (2002) puts it crisply with regards to the surveillance of communication: "Technology has seen to it that our lives are more convenient and more secure, but the price we pay is privacy. In the end it comes down to whether we trust those in authority who are now able to see our telephone and e-mail records."

The notion of agency underlying the concept of biotechnology clearly presupposes a strong agent, which in turn would make the term 'biotechnology' a mere name for the various acts of this agent. Thus, this chapter could already draw to its close, if matters were not somewhat opaque.

For in the course of our investigation, biotechnology has appeared as an institution of reference and determination, resulting from a deliberate enterprise of complex interaction between individual agents bound by the particular discourses and modes of being of that institution, at the same time subjects and objects. Thus, while biotechnology is formed by a determinate number of individual acts, as a concept, it supersedes these acts, encompassing more than their actual effects. In particular, it is striking that individual agents would conform with distinct discourses, using terms and concepts reiterating the institutional rhetoric. The martial language in terms of disease and nature in general, the stories of success, the political creeds and the economic tales of hope are all examples of such discourses.

It is important to stress that I use all these terms descriptively, because they represent standardised forms of communication about the way in which biotechnology is understood, employed and regulated. In this discourse, key words are e.g. "potentiality", "growth", "knowledge" and "innovation", but not "limitation", "stagnation", "ignorance" or "repetition". It is at its core optimistic, which follows from the view on agency it rests upon and further strengthens. There are also critics of this development, whose language, not surprisingly, is characterised by pessimistic terms. It seems difficult, then, to find a descriptive way of speaking about biotechnology, which also became quite clear in our survey of COE documents.

What I find interesting about these standardised forms of rhetoric is the fact *that* they occur, because they demonstrate how a particular concept can shape individual acts e.g. of communication, reinforcing the shaping concept. It could be argued that participants in this discourse do have to comply with the standard if they want to sell products, obtain funds or politically promote the field to various constituencies. Thus, the standard would immediately establish a level of mutual comprehension and assumed trustworthiness, benefiting the interests of those involved and also implying that certain things would not be said or only so in a very specific way.

Consequently, the individual limits its freedom of speech freely or subconsciously in order to obtain an envisioned good. The compliance with such a standard also illustrates the power and, more importantly, the confinement of the concept, somehow existing apart from the individual agents, perhaps as an idea. Figuratively, the concept and the language it entails is in the room, waiting to be grasped. The language of the individual is in this regard depersonalised, framed by institutional requirements which are enforced by those accepting them. It is true that language in general possesses an impersonal aspect insofar as it is defined by a group of speakers with inherent strategies of normalisation preceding the individual's acquisition of its own language, the so-called *mother* tongue[393].

Using Kant's notion of sensus communis, Heath underlines this public aspect of language, indispensably placing the other in our language by which we understand ourselves and our context: "In this way language is public and shapes the content of the purposeness of consciousness. Of course, I am not the consciousness of another, yet their linguistic moment is open to me in the use of language."[394] At the same time, I think it is important to see that language is also empowering the individual, since it allows it to express itself in its speech. Indeed, one's manner of speaking is considered inherently personal, because it is an original combination of standard elements. Language is, thus, inherently ambiguous in its ultimate privacy and total publicity.

Now, the compliance with a particular rhetoric style increases the otherwise rarely felt[395] impersonal aspect of language, even if this submission itself can be interpreted as an expression of one's personality. It is, in other words, a loss of originality. This depersonalisation of language is complemented with the depersonalisation of the human act that we encountered when discussing the concept of technique. Biotechnology relies on a variety of techniques, i.e. human acts that will be repeated in the same way regardless of the actual agent, e.g. when isolating a protein or cloning a cell.

This process is indirectly addressed by the European Commission when it maintains that "[a]s such biotechnology cannot be considered as an industrial sector but rather as a set of technologies developed in the field of life sciences. Its application

---

[393]The use of this term is sexist, of course, because it maintains that the mother will be closest to the child, which is not necessarily so. Furthermore, it expects the child to learn the language spoken by its mother, which might not be her native tongue, though. Finally, this term does not respect instances in which a child grows up without parents.

[394] Heath (2001), 180.

[395]I am aware that this is as an assertion. It seems, though, that the human individual in general sees its own language as an utter part of itself and as an appropriate tool for communicating its thinking and feeling to others. Therefore, lack of command of language is experienced as estrangement or powerlessness, precisely because humans want to be able to express themselves.

span over a number of other industrial or service sectors, and agriculture."[396] In this quote, the term 'technologies' could be replaced with 'technique' in the sense it is used in this book, by which it should be clear that the development of distinct scientific techniques allows for application in various fields, precisely because they are not tied to a particular person. After all, what is applied are the biotechnological techniques developed by humans to be copied by other humans in their respective context.

Regarding biotechnology as a set of techniques (or 'technologies' in the terminology of the Commission) to be applied also indicates that it is an entity containing these techniques before the actual application. Simply put: in order to apply something, it must exist. While techniques as archetypical human acts require an agent performing them in order for them to be - the act first *is* when being enacted - the idea of application nevertheless presumes that they somehow *are* a priori, at least contained as a knowledge about them. Usually, such knowledge can be stored in the minds of those having learned the skills of technique or in some kind of medium, but both instances do not yet clarify fully where precisely the techniques of biotechnology are located whilst not actually being applied, unless biotechnology indeed is understood as an independent entity somehow capable of storing such information.

It is against this background that the term 'biotechnology' is being used in the active mode. Taking the place of the subject in a sentence is, of course, not the same as having it in the reality the sentence seeks to express, but its position is a further indication that biotechnology increasingly is understood as more than a mere name.

As an institution with such characteristics, biotechnology can be compared with an organisation in that it is "neither an individual nor a total social system. It is a subunit of the larger society, comprising individuals in various roles and authorized by the larger society to function for specific, often narrowly defined purposes. (...)"[397]

Some, like Mark Bovens, would define 'organisation' in even broader terms, speaking of it "as soon as two or more people consciously and deliberately cooperate in order to achieve a certain end."[398] Once such organisations are large in scale, bureaucratic and formally established, Bovens and others specify them as *complex organisations* that in the course of the 20th century often have replaced "independently

---

[396]Commission of the European Communities (2001), 97.

[397] Spencer, Mills, Rorty, and Werhane (2000), 25.

[398] Bovens (1998), 10.

operating natural individuals. They have become 'corporate actors.'[399] This has been possible, because of their increase in sheer numbers and in the social significance with which they are accredited. Bovens does admit that it can be cumbersome to prove this thesis, but he interprets empirical data as at least indicative of its validity: there have been very substantial increases in the number of corporate bodies[400], bureaucratic organisations[401], law suits to which a complex organisation is party[402] and in the attention given to them in the media[403]. This has, apparently, led to a new assymetry in society, "no longer between individuals, but between two fundamentally different types of actors: people and complex organisations. (...) Assymetrical or not, modern western societies stand or fall with complex organisations."[404] If (complex) organisations indeed are the key to understanding modern societal interaction, they must have some ethically relevant status, since ethics basically deals with questions of agency.

Edward Spencer et al. point out that "although organizations are not individuals, and therefore are not moral persons, they can be meaningfully said to be moral agents in several senses."[405] Accordingly, there are four aspects that would warrant this assertion of moral agency: organisations set goals[406], they act in the sense that

---

[399]Ibid.,11.

[400]In the USA, the increase has been more than fivefold from 1917 to 1969, "many times more quickly than the population." In the Netherlands, the increase is tenfold after forty years, with 194,000 corporations in 1994. Ibid., 13.

[401]Referring to various studies, Bovens concludes that a strong bureaucratisation has taken place, which also can be inferred from the very high number of wage earners and salaried employees, now mounting to 80 or 90 per cent in most OECD countries. Ibid.,14.

[402]Between half and three quarters of trials now involve complex organisations. Ibid, 14.

[403]This is, perhaps, the weakest point made by Bovens, since he merely refers to the front pages of the *New York Times* and the Dutch *Algemeen (NRC) Handelsblad*, with the considerable decline in their reports on individuals. Ibid., 15.

[404]Bovens (1998), 19.

[405] Spencer,Mills, Rorty, and Werhane (2000), 25.

[406]"These goals are often specified in mission statements, delineated in charters, or defined by the founding arrangements that constitute the organization as a corporate entity." Ibid., 26.

individuals execute collective decisions[407], they are evaluated normatively by others[408] and held accountable[409]. Bovens argues a little indecisively in this regard: since organisations e.g. lack a conscience in the human sense of the word, they are neither moral persons, nor moral entities in all respects. Nevertheless, the idea of 'moral responsibility' would pertain to complex organisations and they can be analysed with at least certain moral categories.[410]

There is much to said for regarding biotechnology as a subunit in this sense, but it differs from organisations by the absence of clear boundaries. A formal organisation is e.g. defined as an impersonal decision-making structure, "where choices of the organization are executed in terms of its mission, goals, and culture. Organizational decisions are social decisions made by individuals on behalf of the organization, and the decisions are attributable to the organization rather than to the individual organizational decision maker."[411] This raises the issue of responsibility, for the more decisions result from interactions of different levels and units of an organisation, the lesser can they be linked to particular persons. Responsibility, thus, becomes fragmented.[412]

Organisational features appear in segments of the biotechnological sector, too. Political bodies express the will of the electorate, pressure groups or governments, and companies seek to comply with the interest of their boards, shareholders or 'markets'[413].

---

[407]"...the actions of an organization are often the result of collective, not individual, decision making. The policies of an organization arise from deliberation between possible courses of action to meet organizational goals. Courses (...) selected (...) involve both their external (...) and (...) internal constituencies." Ibid.

[408]"They are judged to be morally acceptable or unacceptable by other organizations with which they interact, by the individuals who come in contact with them (...), and also by the larger society (...)"Ibid.

[409]"Organizations are held accountable on all the poles of normative evaluation: as agents, on the nature of their actions, and by the effects of their actions." Ibid.

[410] Bovens (1998), 57.

[411] Spencer, Mills, Rorty, and Werhane (2000), 21 with reference to H. Simon: Administrative Behavior. 2nd ed. Free Press: New York 1965, 9.

[412]Cf. Meulenbergs (2000), 32.

[413]This word is, perhaps, the most vague in this context. It does refer to the interaction of demand, supply, and competition, but it also covers public regulation. At the same time, it seems to suppose a certain uniformity of individual behaviour, allowing to group and influence, if not manipulate it.

Reflecting the impersonal structure of decision making, criticism is in the field usually launched against organizations or the whole field, but seldom against a particular researcher, administrative director or clerk. The structure of biotechnology is formed by other agents than organisations, though, which makes it somewhat less tangible. Notwithstanding their reliance on impersonal administration, organisations still have identifiable points of reference in terms of communication or location, which are further stressed by the development of a corporate identity with logos, uniforms etc. They are visible.

The structure of biotechnology does have loci of such visibility in texts[414], human gatherings[415] and edifices[416]. It also has symbols representing it, taken from its basic material such as the DNA double helix or from the range of its products, e.g. the ubiquitous sheep 'Dolly'. At the same time, it seems present without such ostensible elements, e.g. when a particular language is being used or when thought about in the minds of those engaged in its development. This is, to some extent, also true for organisations, since "those working within or for the organization act with those purposes, statements, and goals in mind."[417]

Still, biotechnology exceeds by far the boundaries of organisations that contribute to its formation, because it lacks their clarity in terms of targets, choices, and participants. In other words, the interaction taking place in biotechnology is less coordinated than in an organisation, usually having some clear structure of goal, command and execution. As an institution, then, biotechnology bears in certain respects resemblance to an organisation, but among the differences, the most important might indeed be that *institutional* adaptation and conformity emerge in spite of the absence of a precisely defined power.

In an organisation, submission to particular practices in words and deeds is a contractual requirement for employees, which they might accept in order to obtain

---

[414]Whenever the word occurs in writing, e.g. in a corporate or juridical document, the reality of the concept it signifies is asserted, but its existence not*necessarily* ascertained by the same token, which I have pointed out earlier.

[415]The discussion of biotechnological issues at a world congress of bioethics or in a business meeting of a pharmaceutical company are such instances.

[416]A biology laboratory, an industrial plant or an infertility treatment clinic manifest the presence of biotechnology, but they can only do so, because these edifices are used*and* perceived in this way.

[417] Spencer, Mills, Rorty, and Werhane (2000), 27.

certain goods. In order to ensure compliance with it, the means of control plays a vital role, since "...oversight, monitoring, or reporting relationships serve a positive function: they set the context by which the organization or personnel can measure themselves. In this way, controls establish the context for autonomy and encourage agents to factor the good of others and the organization into their decision making. The virtues associated with positive understanding of control are accountability, trust, and empowerment."[418] In organisational theory, then, submission to the structures of control is regarded as liberating and stimulating for the individual, because it provides standards and limits against which the individual can understand and perform. This view is not altogether new, since it reflects the idea that true human freedom does not lie in arbitrariness, but in the freedom to excel in virtue.[419]

By contrast, the submission to the structures of determination and reference marking institutions occurs without someone demanding it specifically. The process of institution is, thus, effectuated by the individual who in instituting, e.g. in adopting a specific rhetoric, is being institutionalised. This may result from structures of power, but the act of conformity is complex, because it is instigated by an anticipation of the judgement of others, with the psychological notion of 'belonging' as a central motif. The exercise of power is in this respect discreet, because it is communicated through societal symbols rather than individual commands, which means that the individual disciplines itself. The institution does not even require a precise seat of power, then, because the individual subjects itself to the structures of domination that promise empowerment. In this sense, it is not really clear *who* wants biotechnology to happen the way it does.

Moreover, the various political instruments designed to stimulate and limit biotechnology, i.e. to regulate it, demonstrate that its present state does not conjunct with the will expressed by public representatives. Differently put, the desire for change must be fed by a discontent with the actuality of biotechnology, which makes it appear

---

[418]Boyle, DuBose, Ellingson, Guinn, and McCurdy (2001), 190.

[419] Pinckaers (1995), 355, uses the example of a piano in order to illustrate real freedom: "Of course anyone is free to bang out notes haphazardly on the piano, as the fancy strikes him. But this is a rudimentary, savage sort of freedom. It cloaks an incapacity to play even the simplest pieces accurately and well. On the other hand, the person who really possesses the art of playing the piano has acquired a new freedom. (...) His musical freedom could be described as the gradually acquired ability to execute works of his choice with perfection. It is based on natural dispositions and a talent developed and stabilized by means of regular, progressive exercises, or properly speaking, a *habitus*."

as at least somehow involuntary. This discrepancy can in part be explained by the heterogeneity of interests instigating the vast number of individual acts that by their interaction at various societal levels shape biotechnology. The result is far less predictable than the underlying notion of agency would imply, which is characteristic for institutions lacking distinct chains of cause. Indeed, it seems that contemporary societies have a general tendency to develop by series of coincidences rather than planning. While the idea of strong agency still is maintained, the surprising turns of electorates, economies and events bear witness to a world in which individual agents that might have thought of themselves as masters of evolution have become puppets of influence.

From all this, it follows that biotechnology is a concept taking on the character of an institution, based on a number of ideals e.g. about human nature and its abilities. Therefore, it is more than a name, but it seems neither adequate to characterise it with the classical term 'nature', for its heterogeneity turns it into anything but a clearly defined entity. Instead, our reflection has led us to a mode of analysis that some equate with a postmodern approach. Consequently, it is appropriate to complement our investigation with the postmodern critique of the notions of self and technology employed in modern interpretations.

CHAPTER 4

BIOTECHNOLOGY BEYOND MODERNITY?

There seems little doubt that contemporary societies can be understood in so-called 'post-modern' terms. This view is now so common that it is nonchalantly referred to in newspapers. Allegedly, there are no longer common values or collective interpretations of reality, yet this state of nihilism has resulted in the return of philosophy.[420] There is also some agreement that "the postmodern should not be understood as something purely and simply distinguished from the modern, but rather (...) as subsisting in the heart of the modern and offering the labor that will shape its representation."[421] The postmodern project is an attempt, then, to rescue modernity by deconstructing its fallacies and ideosyncracies. We encountered an example of this attitude when reflecting on the role of the human body in medicine: Levin and Solomon suggested that in postmodern medicine, the states of being are respected as well as the processes forming them. The mechanistic perception of the body can, thus, still provide images that are helpful for understanding its functions, but it needs to be balanced by the system models explaining the complexity of the occurrence of e.g. disease in terms other than mere causality.[422]

Furthermore, at the fundamental level, any postmodern criticism of modern rationality still relies on some form of rational discourse, in which insights and meaningful communication are assumed. As far as one would identify such a discourse with the modern project, postmodernity demonstrates its dependence on certain modern

---

[420]"Si elle a davantage droit de cité, c'est peut-être que les grands récits, les grandes légendes ont disparu, que le christianisme ne fait plus recette comme autrefois, ni le marxisme, ni la psychanalyse, et que l'on se retrouve face à soi-même, sans religion, sans dieu, sans gourou, sans utopie et sans avenir. Dans ce climat nihiliste, la philosophie peut fonctionner comme une alternative au sens suprême." Onfray (2002).

[421]Melehy (1997), 10.

[422]Levin and Solomon (1990), 525f.

125

structures.

In modernity, the great narratives prevailed, of which that of modernity itself perhaps was the most significant, maintaining that the world could be understood and shaped by the human subject that lived in societies with high degrees of categorisation and public ideology. As a result, the individual was institutionalised in terms of its identity, behaviour, thinking and space. Its *identity* was shaped in relation to its gender and other social functions, e.g. within the family or the nation. In many ways, modern identity was about adapting to envisioned ideals of being, turning the individual into an instrument of external expectations and goals. Self-perception, thus, reflected social framing. The *behaviour* of the individual was likewise formed by explicit or implicit codes of conduct standardising human interaction. Its work would be instituted as a profession, granting public recognition in return for submission to the professional stuctures. As far as *thinking* is concerned, the mind of the individual was in the modern outlook discovered as the primary space of discipline and punishment. The modern concepts of e.g. agency and autonomy were not only rooted in, but also communicated through human reason. With the strong emphasis on rationality in modernity, individual thought and speech were supposed to conform with its principles, at least when communicated, which was primarily ensured by the means of education. Finally, the *space* of the individual was in several ways deliberately formed: the public space arose through voluntary or coincidental architectural expressions manifesting relations of status, dependence and control. In the actual institutions of modernity, such as hospitals, schools, barracks, religious convents[423] and prisons, the normalisation of the individual took place through positing in standard settings, where personal expressions were limited, if not altogether banned.

These four dimensions of modern humanity have found their most radical interpretation in the dualism of fascism and socialism[424], which can be regarded as the

---

[423]This example may surprise some. It is quite obvious, though, that in the course of the restauration of regimes and the church in the 19th century, church life was highly influenced by such modern modes of being. Seen from outside, it may be quite difficult to see difference between any of these edifices. It was neither unusual to use martial language in describing the objective and strategy of religious life. Even apostolic orders, originally envisioned to gather individuals wishing to respond to actual needs in various settings without any external stability, became increasingly institutionalised.

[424]When speaking of socialism in this regard, I refer to the Soviet systems. It could be argued, of course, that the states that considered themselves socialist in fact were not, at least not in terms of their historical origin, but rather oligarchies founded on state socialism, in which the individual always

stunning finale of modernity: ideally, both societal types were conceived as having abolished class distinction. In these societies, work had become increasingly professionalised and the will of the people was regarded as in total accordance with that of its leadership. While paying tribute to the new human being characterised by physical force and willpower, that same strong subject was belittled in stern public architecture displaying the superiority of collectivism over against non-human nature and human individuality. In combination with industry and education, biotechnology was considered pivotal to the erection of these new societies, in which public health was to be obtained through a variety of means, including eugenics, screening and improved food production.

In view of the fate of the strong subject, it is almost self-evident that postmodernity distances itself from attempt to reconstruct this form of individualism. Therefore, the task does not consist of "reaffirming the subject. We should honor the subject by finding its adequate successors, adequate both in terms of the problem and in relation to the social structure of contemporary society."[425] Likewise, it seems appropriate to examine whether one of its major fields of interaction, namely biotechnology, needs modification in order to make it as contemporary as the subject it presupposes. In this, the anachronism of biotechnology resting upon a subject that has been abandoned could be overcome.

This chapter will begin with a presentation of a postmodern view on the possible successor of this subject, followed by a reflection on one of the major critiques of modern technology, Jacques Ellul, serving as a template for reformulating biotechnology.

1. A POSTMODERN SKETCH OF THE HUMAN SELF

The deconstruction of modern concepts, which is one of the main postmodern enterprises, is dependent upon the very objects it seeks to dismantle. This is not surprising, for any alteration, renovation or demolition can only take place on existing grounds. In the preceding chapter, our reflection on the agent of modern biotechnology led us to the Cartesian thought and in particular to the cogito forming its very core. For postmodern thinkers, this cogito still is a stumbling block, indicating its provoking

---

is subjected to the collective.

[425]Luhmann (1986), 320.

magnitude.

While widely contested, the cogito still is a convenient short hand for the modern understanding of man and the state of anthropocentrism resulting from it. In their introduction to philosophy, which perhaps is much more of a manifesto, Gilles Deleuze and Félix Guattari frequently arrive at this cogito. Discussing the concept of personhood, e.g., they begin with situating this concept. In spite of its having been "created as a concept, it has presuppositions. This is not in the way that one concept presupposes others (for example, "man" presupposes "animal" and "rational"); the presuppositions here are implicit, subjective, and preconceptual, forming an image of thought: everyone knows what thinking means. Everyone can think; everyone wants the truth."[426] The cogito is, thus, dependent upon empiricism that can only be verified subjectively.

Furthermore, the Cartesian concept has three components: doubting, thinking, and being. As we saw above, doubt is the starting point for the investigation that would lead Descartes to see his own thinking as proof of his own existence, for thought requires someone thinking. The concept of the cogito condenses at the 'I' that is shaped by the respective 'I's' performing these three acts. And they are, of course, inseparable, since it is the same self acting. Evidently, this concept does have precursors,but the specific Cartesian contribution lies in "challenging any explicit objective presupposition where every concept refers to other concepts (the rational-animal man, for example). It demands only a prephilosophical understanding, that is, implicit and subjective presuppositions: everyone knows what thinking, being, and I mean (one knows by doing it, being it, saying it)."[427] The point is, then, that there is no objective truth in the cogito, because it rests on mere subjectivity. This is not really problematic, for a "concept always has the truth that falls to it as a function of the conditions of its creation."[428] When a concept is being used, it is created anew and, thus, rooted in underlying interests that frame the value and role attributed to it. This also occurs when former concepts are re-introduced into a contemporary setting.

While it might seem anachronistic, Deleuze and Guattari argue that "one can still be a Platonist, Cartesian, or Kantian today, (...) because one is justified in thinking that their concepts can be reactivated in our problems and inspire those concepts that need

---

[426]Deleuze and Guattari (1994), 61.

[427]Ibid., 26.

[428]Ibid., 27.

to be created."[429] Differently put, the individual's thinking can imitate that of others in order to find an appropriate response to a current problem. Such creation of new concepts prevents some concepts from appearing and makes others disappear, but a concept is "never valued by reference to what it prevents: it is valued for its incomparable position and its own creation."[430] By creating a concept, the human self expresses its interest at a given moment and is, therefore, focussed on the purpose it is envisioned to serve. With such an emphasis put on the creativity with which concepts are shaped, it seem that a strong subject is required after all. The emergence of the cogito would in this sense manifest this subject and its ability to centre itself.

Referring to Kant, Deleuze and Guatteri identify a fundamental weakness in Descartes' view on the subject, namely time: "For it is only in time that my undetermined existence is determinable. But I am only determined in time as a passive and phenomenal self, an always affectable, modifiable, and variable self."[431] Since the self is in time, it is changing, receiving rather than shaping. The *I* that through its thinking determines its being, gains existence as "a passive, receptive, and changing *self*, which only represents to itself the activity of *its own* thought. The I and the Self are thus separated by the line of time, which relates them to each other under the condition of a fundamental difference."[432] This self is in this view depicted as becoming, i.e. in the making - floating instead of a static, clearly defined entity with a substance that embodies an eternal idea.

We encounter this thought in Sartre's work, too, e.g. when he defines human life as a long waiting. There is the waiting for the results of the activities we engage in and, more fundamentally, a waiting for ourselves. These forms are in his view not expressions of contingency, because they do not stem from a nervousness seeking to avoid the present, but "from the very nature of the for-itself which 'is' to the extent that it temporalizes itself. Thus it is necessary to consider our life as being made up not only of waitings but of waitings which themselves wait for waitings. There we have the very structure of selfness: to be oneself is to come to oneself."[433] Deleuze's 'line of time' would correspond to Sartre's 'waiting' and the 'I' to the Sartrean 'oneself', both of

---

[429]Ibid., 28.

[430]Ibid., 31.

[431]Ibid.

[432]Ibid., 29.

[433]Sartre (1956), 538.

which underline the uncertainty characterising the indeterminable life of the fragmentarised human 'self' that the substantial human 'I' seeks to determine. Hence, seemingly contradictory components form the cogito. Thinking, I am active. As a being, I am only determinable in time, and thus passive.

This insight is not really that new, though, for as Simon Critchley points out, already in classical thinking, e.g. in Aristotle's, we find the idea that "matter (...) persists through the changes that from (*morphë*) imposes upon it. In remembrance of this sense of the subject, one still speaks of a subject matter (*ë hupokeimë hulë, subjecta materia*) as that with which thought deals, the matter of discussion or the subject of a book or painting. Indeed, one immediately here notes the oddity that the word *subject* can also designate an object."[434] As matter, the subject is subject to change and, thus, quite as much an object.

In spite of this inherent tension in the subject, modernity placed its trust in it, to the extent of its divinisation. Consequently, Kant's project is also a discovery of the "modern way of saving transcendence: this is no longer the transcendence of a Something, or of a One higher than everything (contemplation), but that of a Subject to which the field of immanence is only attributed by belonging to a self that necessarily represents such a subject to itself (reflection). The Greek world that belonged to no one increasingly becomes the property of a Christian consciousness."[435] As I have argued above, this movement from logos to immanence is enabled by a trust in human reason further sustained by the step taken from regarding it as divinely willed to declaring its own transcendence: "No longer content with handing over immanence to the transcendent, we want to discharge it, reproduce it, and fabricate itself."[436] The modern state of anthropocentrism is in this sense nothing but perfect secularisation, while still religious in its fascination with transcendence, albeit now in a mere human form. This movement is propelled by human thinking and, thus, the product of the active self, which, ironically, has arrived at the plane of sheer immanence, for it is "no longer immanent to something other than itself."[437] What is left is empiricism, noting events, i.e. "possible worlds as concepts, and other people as expressions of possible worlds or conceptual personae (...) Empiricism knows only events and other people and

---

[434]Critchley (1996), 13.
[435]Deleuze and Guattari (1994), 46.
[436]Ibid., 47.
[437]Ibid.

is therefore a great creator of concepts. Its force beings from the moment it defines the subject: a *habitus*, a habit, nothing but a habit in a field of immanence, the habit of saying I."[438] The assertion of self, strong as it might seem, is nothing than a habit, then, which we take on while surrounded by the illusions marking our existence: the illusions of transcendence (which actually is immanence), universality (which does not explain, but needs explanation because of its immanence); eternity (forgetting that concepts must be created, and thus depend on time), and discursiveness (confusing propositions with concepts).[439] In other words, the actual individual is always living in time and, therefore, shaped and stratified.

Moreover, this "passive self (...) necessarily represents its own thinking activity to itself as an Other (*Autre*) that affects it. This is not another subject but rather the subject who becomes an other. (...) A new syntax, with other ordinates, with other zones of in-discernability, secured first by the schema and then by the affection of self by self [*soi par soi*], makes the "I" and the "Self" *inseparable*."[440]

Simply put, the concept of an affirming self includes an inherent estrangement that results from the assertion of security parred with an uncertainty stemming from the influences forming the subject. We have encountered precisely this tension in our analysis of European attitudes towards biotechnology: on the one hand, there is the concept of a strong agent instigating processes in living material, ranging from microorganisms to humans, possessing a clarity of mind and a strong sense of mastery. On the other, there is the realisation that the very same subject is endangered by possible misuse of these procedures and far less able to control the processes it initiates, because these are so utterly dependent upon the often indeterminable cooperation of the material dealt with. In this, the alleged master is also a slave to the unpredictability of behaviour found in human and non-human forms of being, be it in the global economy in the case of the first or in the petri dish in the case of the latter.

According to the analysis of Michel Foucault, the human self is likewise objectified by diverse influences, which he understands mainly in terms of power. His focus is steered by the insight that "while the human subject is placed in relations of production and of signification, he is equally placed in power relations that are very

---

[438]Ibid., 48.

[439]Ibid., 49-50.

[440]Ibid., 32.

complex."[441] This power is especially problematic when expressed politically, i.e. in society. For Foucault, the "relationship between rationalization and excesses of political power is evident. And we should not wait for bureaucracy or concentration camps to recognize this existence of such relations."[442] This power is exercised through institutions, groups, elites and classes as well as in a particular form of power, i.e. as a technique. Applying itself to the everyday life of the individual, this form of power "categorizes the individual, marks him by his own individuality, attaches him to his own identity, imposes a law of truth on him that he must recognize and others have to recognize in him. It is a form of power that makes individuals subjects.

There are two meanings of the word "subject": subject to someone else by control and dependence, and tied to his own identity by a conscience or self-knowledge. Both meanings suggest a form of power that subjugates and makes subject to."[443] Hence, the individual has to struggle socially against this power in the forms of ethnic, social and religious domination, the exploitation separating individuals from the production of their work and the ways in which the individual is subjected to itself, and thus, to others. This struggle is difficult, because it is faced with three forms of relationships, namely of power, communication and objective capacity, i.e. the inherent abilities of the body or the external instruments relayed to it. These relations interact

"and use each other mutually as means to an end. The application of objective capacities in their most elementary form implies relationships of communication (whether in the form of previously acquired information or of shared work), it is tied also to power relations (whether they consist of obligatory tasks, of gestures imposed by tradition or apprenticeship, or subdivisions or the more or less obligatory distribution of labor). Relationships of communication imply goal-directed activities (even if only the correct putting into operation of directed elements of meaning) and, by modifying the field of information between partners, produce effects of power. Power relations are exercised, to an exceedingly important extent, through the production and exchange of signs; and they are scarcely separable from goal-directed activities that permit the exercise of a power (such as training techniques, processes of domination, the means by which obedience is obtained), or that, to enable them to operate, call on relations of power (the division of labor and the hierarchy of tasks)."[444]

---

[441]Foucault (2001), 327.
[442]Ibid., 328.
[443]Ibid., 331.
[444]Ibid., 338.

In condensed form, this quotation contains Foucault's analysis of the situation in which the modern self is entangled. It entails that even the seemingly simple act of work actually is an instance of highly developed stratification, within which the communication of learning and cooperation always takes place in settings ruled by explicit (e.g hierarchy) or implicit (e.g. tradition) domination. The individual participating in this diverse exercise of power might not be fully aware of its role, indeed, the very nature of such power is that there is no clearly defined channels. This power does not need violence, although it can make use of it; instead, "[i]t operates on the field of possibilities in which the behavior of active subjects is able to inscribe itself. It is a set of actions on possible actions; it incites, it induces, it seduces, it makes easier or more difficult; it releases or contrives, makes more probable or less; in the extreme, it constrains or forbids absolutely, but it is always a way of acting upon one or more acting subjects by virtue of their acting or being capable of action. A set of actions upon other actions."[445] Since the act of an individual reflects the interplay between personal capacities, communication, and power relations, it further strengthens the exercise of power since it embodies the strategies that form it.

This process relies on certain rationalities, in particular on the ability to enter into people's minds. The whole point of this form of power is to dominate by letting the individual simultaneously perform the task of the executioner as well as the victim, enabled by the "general system of oversight and confinement [that] penetrates all layers of society, taking forms that go from the great prisons built on the panopticon model to the charitable societies, and that find their points of application not only among the delinquents, but among abandoned children, orphans, apprentices, high school students, workers, and so on."[446] By incorporating the demands of power, the self subjects itself, substituting its individuality with the conformity that is expected. As a result, the self is enslaved without perhaps even noticing it.

   In Foucault's view, it is insufficient to identify this process solely with the Enlightenment. Rather, it is deeply rooted in Christianity and the pastoral power of its church, having evolved from four features that characterises this pastoral power. Firstly, its primary aim is the salvation of the individual; secondly, it is a power that is prepared to sacrifice itself for the flock, as in martyrdom e.g.; Thirdly, its concern is the whole community as well as the individual during its entire life; And finally, it explores

---

[445]Ibid., 341.
[446]Foucault (1997), 32.

the mind and soul of the individual in order to know and direct the conscience.[447] The functions of this particular power has spread to the modern state, in which transcendence is replaced with immanence, the exercise of power widened by including public and private institutions and structures and the knowledge of the individual parallelled with the desire to obtain knowledge of the entire population.[448] These movements have resulted in "the disciplining of societies in Europe since the eighteenth century (...)" While this development has not been completely successful in its taming of the human subject, it has a least led to "an increasingly controlled, more rational, and economic process of adjustment (...) between productive activities, communications networks, and the play of power relations."[449] We have encountered this connection between modern and theological concepts before and Foucault's contribution is in my opinion further support for identifying the trust in human rationality as pivotal to this development.

As far as biotechnology is concerned, I have argued above that it may be understood as an institution that collects diverse practices without a clear notion of agency or goal. Hence, there are similarities between the current notion of biotechnology and the movement described above. In particular, the inter-dependence between objective capacity, communication and power also marks biotechnology and the hopes attached to it. Drawing a parallel between the success of the hygienic society, which supports Foucault's analysis, and the future role of genetic information, Henk ten Have deems that

> "[t]he ideals of value-neutrality of clinical genetics and of personal responsibility for health, prevailing in current bioethical debate, may indeed generate a situation where the availability of genetic information in itself produces its wide-spread application. In this view, human beings in the next millennium will be dominated by predictive knowledge of their genome and driven by new norms in interpersonal behaviour. Such assumption is not unrealistic since we have witnessed a similar change in normative behaviour patterns at the close of the last century (...) The new norms of a healthy, regular, and disciplined conduct passed into domestic life; the strategy succeeded in having the norms internalized. Hygienism thus produced a new behaviour patters in the general population."[450]

---

[447]Foucault (2001), 333.
[448]Ibid., 334.
[449]Ibid., 339.
[450]ten Have (2001c), 363.

Such a society could, of course, only emerge because of social practices that made the individual adapt to the standards that had entered its mind. It is, therefore, not surprising that Foucault considers these mechanisms strong entities, "blocks, in which the deployment of technical capacities, the game of communications, and the relationships of power are adjusted to one another according to considered formulae, constitute what one might call, enlarging a little the sense of the word, "disciplines."[451] In analysing this interplay, Foucault accepts that one becomes the object of power relation, without subjecting oneself to power itself.

As we have seen, the act of technique requires subjecting oneself to the requirements that are set by the objects and the environment within which we encounter them. Furthermore, technology forms a discourse framing the discussion of societal development. There are, e.g., particular ways of interpreting biotechnology as the primary instrument of progress, by which research, politics and the industry are influenced, if not altogether marked. It might be, then, that this development forms society in a particular way. In this regard, Jacques Ellul's reflections on the role of technology in general may illuminate this interaction, assisting our quest to determine the precise function of the human subject in the technological societies.

2. TECHNOLOGY AS BLUFF

Jacques Ellul reflects on the relation between technology and the modern societies in which it emerges. At its core, his position is marked by a great reluctance to accept technology as a major force of societal development, in particular because of its destructive tendencies, veiled by a rhetoric saluting the progress granted by new techniques. To his mind, contemporary society is entangled in a technological bluff, which "consists essentially of rearranging everything in terms of technical progress, which with prodigious diversification offers us in every direction such varied possibilities that we can imagine nothing else."[452]

Having presented this thesis in 1986, with the English translation published in 1990, Ellul's contribution could seem irrelevant with regards to our discussion of biotechnology for two reasons: firstly, its at times overly critical tone and vivid fear of

---

[451]Foucault (2001), 339.
[452]Ellul (1990), xvi.

crisis, in particular in terms of pollution[453], over-consumption, 'star wars', nuclear disaster, terrorism[454] and 'computer'domination[455], are deeply rooted in an early 1980s perception. Those were days of transition, with a major decline, if not fall of the traditional industries in Europe, featuring widespread closings of steel mills, coal mines and shipyards. Ellul's indecisiveness in terms of finding the appropriate attribute for society - is it the media, space, nuclear, consumer or education society?[456] - further illustrates this transitional character. The problems he diagnosed then have not vanished since, but the world of 2003 is significantly different in its political settings and public opinion.

Secondly, and somewhat related, technology has also changed drastically since that time, with major developments in biotechnology first occurring in the course of the late 1990s. As we have seen, much political attention is now being given to our field, which was not the case in the 1980s, when it was still believed that the innovation leap from the 1950s to the 1980s would continue, providing "increasingly powerful and versatile types of equipment"[457] in every field. While Ellul thought he made a mistake when assuming in 1977 that particular areas would stabilise once certain degrees of perfection and efficiency had been reached[458], various fields have shown such

---

[453]The reference to acid rain is in this regard typical: "Rainwater has sometimes been no less acid than citrus juice. Studies have shown that the situation in Norway is catastrophic in this field. Fish and forests are affected and corrosion is reported. But the clouds that are charged with acid come from places at which attempts have been made to solve the problem of smoke pollution." Ibid., 59.

[454]When speaking of terrorism, he is thinking in terms of the attacks that so deeply provoked e.g. Germany and Italy in the 1970s and 1980s. Cf. e.g. 201: "But the nihilism which derives from the philosophy of the absurd can also affect those who ar not philosophers or artists, leading some to suicide (...) and others to terrorism (...) My point is simply that this philosophy has penetrated much more deeply than we think and created a climate in society as a whole in which terrorism could develop. Nor must we forget that some terrorists (e.g., the Baader gang) were in fact intellectuals."

[455]Ibid., 5: "There can certainly be no doubt that our society is characterized above all by the computer, and that atomic energy, important though it is, is not what gives meaning to our world." He regards space already as a rival for the computer, though: "Space today involves a number of such new and decisive enterprises that one might say quite definitely that our world is just as much characterized by the opening up and conquest of space as by the computer."

[456]Ibid., 13. He opts then for the society of progress as a tentative term.

[457]Ibid., 1.

[458]Ibid., 1. In 1977, he published Technological Society. Cf. Ellul (1980).

saturation.[459] It is telling, therefore, that in contemporary societies, technology as such is no longer seen as promotor of progress, *bio*technology is. It is therefore not surprising that in his book, biotechnology is being dealt with marginally, while the techniques in transportation, data processing, automatic production and armery still play dominant roles.

In spite of these limitations given by its historical context and the ideosyncracies clearly marking his work[460], Ellul's position is nevertheless a most valuable example of how modern technology can be assessed critically. In the following, I shall use Ellul's main thoughts on the technological bluff as a critique on three of the features that we identified as pivotal to biotechnology: the human mastery of change, the ambivalence of the new techniques and, most importantly, the dominating role of technology.

2.1. Human mastery

The primary agent of biotechnology is the human being instigating changes in the objects he has chosen. As discussed in the second chapter, this underlying view on agency is clearly focussed on a strong agent mastering techniques in order to achieve envisioned goals. Simply put, it is assumed that by virtue of their inherent freedom, humans can do what they will, thus actualising their potentiality. This view is also reflected in Ellul's definition of mastery as "an ability to dispose *at will* one's potential."[461] In Western societies, there is a tendency, then, to propagate the free

---

[459]In transportation, there has been no substantial development in terms of supersonic aeroplanes or magnetic gliding trains. High speed rails services have now widely accepted standards of 200 km/h (or 125 mph), 250 km/h, 280 km/h or 300 km/h, marking levels of different cost-benefit rationales. In maritime transport, nuclear power has not been developed further as an alternative to oil. Increasingly, nuclear power plants have been closed or are at least scheduled to do so, while other sources of energy are exploited, such as wind. Building techniques have not changed significantly and neither have e.g. kitchen utensils.

[460]Certain things he simply does not like, e.g. motorcycles: "Motorcyclists take pleasure in their engines and the pleasure is doubled if they make the maximum noise." Ibid., 75. A motorcyclist myself, I can assure you it is not, mostly because one does not hear one's own engine. He also has a high disregard of homosexuality, for "what was once regarded as an aberration, like homosexuality, will be accepted."127 This is for him a sign of total permissiveness that allegedly only embarrasses if it is destructive. Ibid., note 7. It is a pity that his view on gay life is not quite as nuanced as that presented e.g. by the Catholic magisterium in its most recent Catechism.

[461]Ibid., 157

expression of human potentiality without repression, because it is the hallmark of modern self-understanding.

Ellul is not rejecting the idea that possibility and freedom are pivotal to human existence, on the contrary, he stresses that

> "[t]here is no individual, no human being, no self, if there is no freedom, no possibility. It is no good living if there is no margin of freedom on which the self can constitute itself. Conversely, freedom is not real unless the self comes up against a necessity or a group of necessities. The play between these two realities is what makes human existence possible. We are caught in a web of determinations, but we are made so as to control and utilize them and to achieve freedom in this way. The self is already itself (necessity) but it has also to become itself (possibility). A self that is without possibility is desperate, and so is a self without necessity."[462]

The point is, then, that human existence is not about pure possibility, but about the dialectical relation between freedom and its boundaries, which in our context could be applied to the determinations of technology and individual human freedom. It is in solving this tension that the self constitutes itself.

Insofar techniques are specific types of voluntary acts, their very concept presupposes this potentiality, because they are designs that are meant to be replicated by other human agents. Since the techniques of biotechnology usually deal with the modification or creation of organisms, they carry a silent promise of humans having the power to change their environment, i.e. to culture nature. This mirrors the general function of the technical culture, which according to Ellul provides us with the knowledge and know-how to exercise mastery over nature and to control those humans with whom we need to interact.[463]

As we have encountered earlier, humanity is thus regarded as the measure of all things, which is the state of anthropocentrism so intimately linked with the concept of modernity. Ellul also identifies certain theological roots of this development, albeit in more general terms. With humans seen as creation awaiting its full realisation, they have the possibility, if not duty, to make it perfect in accordance with divine will. It is this goal that technique is to serve.[464]

---

[462]Ibid., 217

[463]Ibid., 139

[464]Ibid., 128

He thus infers to the theological interpretation of creation as divinely willed, which means that human nature likewise is an expression of this will and, therefore, legitimately can use its inherent potential to develop itself and nature further. It is apparent that the new discoveries, e.g. in the biotechnological sector, have resulted in an enthusiasm about the human ability to overcome socially and naturally conditioned inequities, in particular the traditional limitations set by society, morality and the body. Perhaps, we are indeed "embarking on a new Odyssey."[465]

While the new techniques promise mastery and control, the technical growth allied with them is strangely beyond their reach, for once having embarked, human freedom appears as rather confined. It is for instance difficult to prohibit advances in biomedical research and in general, it seems that biotechnology as a common enterprise cannot be stopped, especially because of the variety of interests at a stake and the lack of distinct chains of cause. Paradoxically, then, we are bound by the decisions we have taken, even those that we thought would express our freedom in a paramount way.

Ellul, thus, sees an *absolute* determinism at work, which is "the source of the basic despair of modern humanity. This is the key to it. We despair because we can do nothing and we are vaguely aware of this even though we do not know it."[466] What set out as an enterprise of human freedom, then, has in Ellul's opinion turned into the opposite, now enslaving the agent once considering itself master, which is a conclusion we arrived at from a different angle. This has made it necessary to replace the thesis of mastery with that of fascination. This subsitute concept entails "exclusive fixation on an object, passionate interest, the impossibility of turning away, a hypnotic obedience, a total lack of awareness, and finally exteriozation of self (either possession or dispossession, according to where one is situated)."[467] In other words, fascination is the unreflected submission to an object binding one irrevocably. This is how Ellul interprets the current state of affairs between the human agent and technology.

Since the polydimensional character of technology makes it sheer impossible to direct it, there is no possible orientation and hence, no mastery.[468] Following the traits laid out earlier, where I described the inherent dialectic of technique as binding the agent using it, the human agent might have lost its ability to return to its original state

---

[465]Ibid., 125
[466]Ibid., 219
[467]Ibid., 323
[468]Ibid., 157

of freedom. Ellul believes that this situation can only be overcome by a non-technical power able to regulate and modulate the technical system.[469] To the extent that Western societies appear as dominated by technology, though, such power may have great difficulties in emerging. Before investigating this issue more deeply, we should first address the ambivalence of techniques that has already begun to take form.

2.2. The ambivalence of the new techniques

Techniques are by their very nature unforeseeable, e.g. in the field of genetic engineering[470], because whatever is intended to bring about good, may result in evil. Hence, "[a]ll technical progress has three kinds of effects: the desired, the foreseen, and the unforeseen."[471] In our present context, the risks of new techniques may be forms of lesser evil taken into account or, indeed, negative results that were neither foreseen nor intended. According to Ellul, we speak of lack of foresight "when one might foresee something but does not."[472] This can be caused by the inherent human inability to fathom all that should be foreseen or by the individual's states of emotion clouding its judgement. In terms of public agencies, Ellul sees inadequate materials and insufficient fundings as examples of such lack of foresight.[473] This notion is contrasted with that of 'unpredictability', referring to those instances "when in spite of every effort future events are obscure and one cannot give the probable course of their development."[474] These are situations in which occurs what we could not have reasonably foreseen. It is this unpredictability that characterises accidents, for we know that they are likely to happen, but not where and when.

Nevertheless, we need to take them into account and Ellul criticises experts for not doing so, e.g. when assessing risks and planning disposal of wastes in conjunction with nuclear energy. This creates a state of uncertainty especially on part of the public, which in his view combined with the ignorance of experts forms the locus of unpredictability.[475] It affects technical development as it relies on the three distinct

---

[469]Ibid., 157
[470]Ibid., 54
[471]Ibid., 61
[472]Ibid., 82
[473]Ibid., 83.
[474]Ibid., 82.
[475]Ibid., 83-84.

phases of technique, invention, innovation and diffusion.

The first two phases are dependent upon the effective application of techniques already in use, without which something new cannot take form. Additionally, invention is also highly influenced by public policies and their limitations. The third depends fully on the capital provided for it within the economy, which raises questions of venture capital and markets. The combination of these factors imply that "there is necessarily unpredictability in technical development."[476] Giving the impression of strength and reliability, the technical system is in fact vulnerable due to its need of specialised subsystems. In order to function, it requires agreement on political and economic organisation, a dependent social structure and little violent opposition. Thus, the technical system does not develop in a pure universe.[477]

This is most certainly so in biotechnology, as we have seen, for without extensive direct or indirect public funding, on which decisions have to be taken, the field could not develop. This does not secure success, however, for even costly research and development may not produce anticipated results due to the complex biochemical reactions in the human body, for instance. Increasing resistance in recipients to certain pharmaceuticals, such as antibiotics, demonstrate that is most difficult to predict the exact effects of biotechnological products. This difficulty stems from the often indeterminable connection between cause and effects inherent in techniques that rely on the co-operation of the organisms used.

It may very well be, then, that the subsequent manifestation of unforeseen risks represent degrees of harm that would have barred the application of a technique had they been known in advance. It is this concern that e.g. marks the COE documents in terms of the protection of the individual being. While it is not assumed, of course, that the new techniques are designed to harm individual humans, they may do so in the course of their application. Hence, the positive and negative effects of techniques are inseparable. Indeed, in view of the impact of technology on traditional cultures, Ellul

---

[476]Ibid., 91. Ellul presents here the following example in this regard: "When the industrial use of nuclear fission coincided with the early stages of the computer, no one could say which would win the market or become the dominant force and shape society. The latter won, but no one could foresee this."

[477]Ibid., 117.

maintains that "technical progress almost always has negative effects on people."[478]

In his opinion, also the increase in life expectancy is ambiguous, for it creates demographic and physiological problems of the lives thus prolonged. As a result, we "live diminished lives and do not have the same vital force. We have to compensate for new deficiencies by artificial procedures that in turn produce other new deficiencies."[479] The basic problem is, then, that while technique brings undisputable benefits with it, it destroys goods already existing.

Ellul does acknowledge, however, that progress has brought about material advances, and that not every creation is equalled by destruction. The point he wishes to make is that one should be hesitant in terms of maintaining that progress indeed has taken place.[480] I would see this as a call for caution, seeking to avoid the normative language in which the results of technology are evaluated as positive *per se*. Hence, "[t]here is no progress that is ever definitive, no progress that is only progress, no progress without a shadow. (...) There is a double play of progress and regress."[481]

This situation is complicated by the fundamental ambivalence of technology. For the human agent is not separated from technology, but already "closely implicated in this technical universe. We are conditioned by it. (...) The use we make of equipment is not decided by spiritual, ethical, autonomous beings, but by people within this universe. Thus this usage is as much the result of human choice as it is of technical determination."[482] Humans are already captured by a system offering them means and ends, situated in the technological universe.

Consequently, assessing its instruments is quite a task, for in order to decide freely on the acceptability of a given practice, one would have to step out of the context on which one is dependent. In this regard, the question remains whether the ambivalence of techniques indeed would make it possible to take every factor into account, adopting an overall standpoint, which in turn might warrant paralysing the technical action due

---

[478]Ibid., 41 He gives two examples: "Roads and truck transport in the Sahara have replaced the caravans of that were the source of wealth for the Tuaregs. The increasingly efficient technicization of seal hunting has deprived the Eskimos of something that was basic to their society."

[479]Ibid., 44, This is a somewhat delicate passage, for Ellul blames progress for resulting in a greater number of old people needing care and the survival of weak infants with a fragile health or handicap.

[480]Ibid., 46

[481]Ibid., 71

[482]Ibid., 37

to the uncertainty it conveys.[483]

The discourse of rationality is often seen as a means of obtaining such a neutral perspective. Yet also at this point, Ellul remains sceptical, because the alleged rationality of technique is merely masking underlying currents of technical exploitation and objectification of things and beings, in fact resulting in chaos, which science in turn is supposed to resolve. The discourse on rationality is in this regard a means of justification, fulfilling the role once possessed by religion.[484]

While acknowledging that technique used to function in a mode of rationality, Ellul sees it now falling into irrationality, contributing nothing of value, as in mere services and data processing, and tending to destruction.[485] Modern technique is in this sense linked to the philosophy of the absurd, he argues, because it produces absurd behaviour and creates absurd economic situations.[486] It is, thus, for him absurd to assume that future needs and situations indeed can be predicted.[487] As regards the economy, the focus on techniques makes it feeble, because it is "neither possible, predictable, rational, nor capable of global organization. (...) It is a remarkable fact that an excess of technique always leads to absurd situations, to an impasse from which it is impossible to see any exit."[488] This is somewhat dramatic language, but I take it as an attempt to reiterate the point already made, namely that by focussing *solely* on technique as a vehicle of societal development, society will become increasingly vulnerable, especially because the economic development at heart is indeterminable. For Ellul, this is worsened by the inherent tendency in technique to produce more of the same sort regardless of its actual value.[489] In view of the burst of the IT market in 2001 and its rather slow recovery, it might be that a too narrow focus on biotechnology as a means of societal progress could lead to similar economic instabilities. At any rate, there *are* determinations of this technical universe "that dictate a certain use."[490] Consequently, the impact of these techniques might be more comprehensive than

---

[483] Ibid., 60
[484] Ibid., 170
[485] Ibid., 72
[486] Ibid., 203
[487] Ibid., 206
[488] Ibid., 212.
[489] Ibid., 263.
[490] Ibid., 37.

envisioned.

Ellul illustrates this by the proletariat, which he regards as having emerged from mechanisation and the division of labour rather than directly from capitalist acts. The capitalists were, thus, only intermediaries starting the process of production that applied existing techniques. The fact that technisation in the USSR led to "the creation of a proletariat at least as unhappy as that of England in 1850"[491] is further proof that techniques do create problems that might be much more complex than anticipated. With this historical reference, Ellul wishes to show that technique is not neutral and in that differing from mere instruments, such as a knife, by inherently having positive and negative consequences regardless of its use.[492] Indeed, Ellul maintains that the atrocities of our age would have not been possible without technique, and even though humans are responsible, and not matter, the fact that the very same people have placed high hopes and emotions in the development of techniques indicate the link between brutal reality and technical growth.[493] The humanistic discourse on anthropocentrism is in this regard a contribution to the bluff masking this truth.

As a result, the negative effects of technology seem not very present in the public discourse, bearing witness to the extent by which societies have been shaped by techniques. This has been made possible due to five paradigms directing this process: Firstly, there is a desire to normalise, e.g. language, in order to obtain compatibility with technical requirements. As discussed above, this normalisation also appears as acts of standardisation in conjunction with the alleged 'globalisation'. Secondly, one senses a general obsession with change, which is the popular form of the myth of progress. I have shown that this positive attitude towards change is not really that surprising, considering that this concept is inherently linked with the human understanding of agency and self. Thirdly, Ellul identifies the desire to foster growth at all costs, which we also detected in the political documents used for our analysis. There is a strong belief in the positive effects that economic growth will bring to society at large, with biotechnology pictured as a major catalyst. Fourthly, there is a widespread fascination with speed and, finally, a tendency to make all judgements dependent upon technique, i.e. not to allow any critical assessment of progress.[494] Even if public awareness of these

---

[491] Ibid., 49.
[492] Ibid., 35.
[493] Ibid., 131.
[494] Ibid., 223-226.

negative effects of technology were to develop, it would face three major obstacles in Ellul's view, namely the influence of the military-industrial complex, the enormous public and private capital at stake, and the fact that damages and dangers are assessed in money only.[495] The public thus feels "bound up to a kind of destiny that is beyond their control and that might be triggered at any time. It is this uncertainty which causes panic (...)"[496]

This observation is supported by the experience of public reaction to new developments in biotechnology, where immediate rejection and a state of alarm form typical attitudes. It is not unusual for humans to react with immediate rejection when they encounter novelties. The general human response of initial rejection, subsequent evaluation and final acceptance is often also present in our field. Such patterns could be broken, however, by a comprehensive information about the precise nature of the techniques and their risks and benefits. What is needed then, would be a true assessment of all the advantages and all disadvantages of a technique, also those that surpass the monetary level. Yet this is unthinkable in view of the irrepressible ambivalence that marks technique.[497]

Still, Ellul propagates foresight taking all essential facts into account "so as to have a response to the worst scenario in each case. We have to weigh what is probably the worst in each situation and then find a solution for it."[498]

In sum, Ellul deems technical progress ambivalent, because it has a price, at each stage leading to more and greater problems than it solves with its largely unforeseeable effects that intertwine harm and beneficence.[499]

2.3. The dominating role of technology

The technological development binds humans by preventing them from bringing it to a halt, reinforced by their using it as a vehicle of their self-understanding that replaces myths and great projects with the foundation of the technical system, in which techniques produce "the most reassuring and innocent ordinariness."[500] The religious

---

[495]Ibid., 75-76.
[496]Ibid., 87.
[497]Ibid., 76.
[498]Ibid., 99.
[499]Ibid., 39.
[500]Ibid., 19.

feeling is transferred to objects, e.g. television, computers, bikes (!) and rockets, that are now regarded as fabulous, powerful, unlimited, ubiquitous, awesome, worthy of sacrifice , i.e. in short as sacred.

While nature used to form the space of religious encounter, the fictional world humans have created, "in which our religious sense incarnates itself[,]"[501] has now replaced it. Obviously, Ellul assumes that humans have an innate religious inclination.

Thus emerge Western societies where technique forms the environment, "the new 'nature' in which we live,"[502] taking on a dominating and systemic character. Society is more than the technical system, still remaining outside of it with institutions that are not 'rigorously' technical and with a general presence of ideologies, history and myths.[503] Techniques have begun to mark these other domains, though, arriving as an unavoidable feature of society: "Nothing at all is beyond its grasp. It is *causa sui*. Ordinary common sense expresses this in the saying that we cannot stop progress. But this popular phrase has now become the last word in all consideration of these phenomena."[504] This has an impact on culture, for it is nurtured by myths and rites of popular creativity, in "a communion of style between life and art."[505] In Ellul's view, technical and computer activity excludes such creativity and spontaneity, though.

Furthermore, technique is not concerned with the meaning of life and cannot itself give insight into new values. It cannot even accept value judgements about itself.[506] Technique radically alters society, then, affecting its fabric predominantly in six ways: firstly, there is an almost total disappearance of ends that have been thought out and clearly conceived (at all levels). There is, thus, no clear direction of societal development. Secondly, human interest in the concrete sense fades and is replaced with entertainment (which is another way of diagnosing disinformation through excess of information[507]). Thirdly, technical progress is in itself equalled with the good. Fourthly, combinations of very complex multiple interests emerge, replacing classical economic, political, and social interests that are now largely outdated. Fifthly, the inability to grasp

---

[501] Ibid., 121.

[502] Jacques Ellul: op.cit, 15.

[503] Ibid., 16.

[504] Ibid., 218.

[505] Ibid., 140.

[506] Ibid., 148.

[507] Ibid., 329.

a situation globally; And finally, an inability to rectify our mistakes by analysing the path we have taken and seeing the factors at work on it.[508] In a nutshell, Ellul thus asserts that the submission to technological progress damages the very notion of agency, turning the individual human being into the object of vague interests, influences and developments that inhibit rather than empower.

What makes this development inevitable is the shaping of people who do not think otherwise, then, and not the development of science or technology as such. Hence, techno-science presupposes "the belief that it is, the pseudo-predictive boasting, and the assuring of people that is in the process of realization."[509] This is facilitated by the bluff that presents the discourse on techniques, namely "technology", as an expression of actual processes, pretending that words signify reality: "There is a bluff here because the effective possibilities are multiplied a hundredfold in such discussions and the negative aspects are radically concealed (...) it transforms a technique of *implicit and unavowed* last resort into a technique of *explicit and avowed* last resort (...) caus[ing] us to live in a world of diversion and illusion"[510] While we approach catastrophe, the bluff secures that the public rests assured of the enormous potential of the technology that will cause it. It is the superexponential growth of technology in finite time, with the production of things that cannot be eliminated, that leads Ellul to believe that psychological disorders will follow from it, contributing to the disintegration of society.[511]

In technological societies, productivity is a question of reducing energy and labour for the same outcome. Thus, an "enterprise is now more productive and competitive the less is employs human labor."[512] This means that in turn, technical progress also entails that certain humans are considered un-exploitable, because they do not have the mental or physical capacity to comply with the standards of constant and high performances.[513] Since growth is considered pivotal to solving socio-economic problems and to be achieved through the competitiveness of higher productivity only, it is the role of

---

[508]Ibid., 242.

[509]Ibid., 400.

[510]Ibid., xvi.

[511]Ibid., 98.

[512]Ibid., 5.

[513]Ibid., 58-59. Ellul's observation has proven to be correct in view of a somewhat constant unemployment base remaining in many Western European countries notwithstanding intermittent economic revivals.

science to ensure this, which means that "science itself has an economic effect."[514] And while the dealings of science grow increasingly complex and incomprehensible, this only seems to increase the positive attitudes and beliefs attached to it.[515] Science has become our only recourse, then, and it provides the means of salvation for the individual and the race, solving the problems of war and pollution.[516]

In early scientism, science was used in order to suppress God and prove his non-existence. This is no longer necessary, for we have arrived at the same point, where God no longer plays any role. Science is divine and God, thus, obsolete.[517]

The duty of the state is accordingly to promote techniques and to engage in their operation.[518] By the same token, the state can legitimise its own power with science and technique rather than religiously or democratically. Science and the state thus enter into a relation of co-dependence, because societal development is regarded as dependent upon research that will develop new techniques.[519] By promoting science and directing it to high levels of technical production and progress, the state intends to achieve its goals, such as good production, balanced budget, good exports, adequate domestic consumption, continued growth.[520]

Apart from deemed inherently valuable, technique is also seen as a major tool for obtaining values such as justice, happiness[521], the reduction of work etc.[522] Technique appears, thus, no longer a tool of economy, but as its driving force.[523] On the other hand, science needs the state since it can only develop with techniques that are beyond the means of strongest corporations. This is, of course, applicable to biotechnology, where the governments are clearly interested in promoting research and development, while the industry emphasises that public funds are needed in order to stimulate the sector. If one of the main goals of the state indeed is to secure its own stability,

---

[514]Ibid., 181.

[515]Ibid., 181.

[516]Ibid., 183.

[517]Ibid., 184.

[518]Ibid., 24.

[519]Ibid., 103.

[520]Ibid., 302.

[521]Happiness not in the classical sense, but as meeting needs, assuring well-being, gaining wealth, culture, knowledge. Ibid., 259.

[522]Ibid., 30.

[523]Ibid., 244.

biotechnology is clearly seen as one important pillar. In terms of the relation between technology and science, we encounter "a strict circle. Science accelerates technical progress and technical progress reacts by making possible new discoveries. Much technical progress is of value in making scientific development possible. And since science is the final justification of our Western world, money devoted to these technical exploits seems to be legitimately spent."[524] In such technological societies, specialisation is required, which necessitates the acquisition of competence.

It is in this context that a new aristocracy appears, the "technocracy", in which technicians add authority to their competence.[525] Like aristocrats were above the law, so technocracy is it today, in that failure or accidents usually are not attributed to technique, but human error.[526] Furthermore, they serve society in such a way that their work seem indispensable: "They are at the center of every organism of management and decision. Armaments, space exploration, medical treatment, communications, information, industry, administrative rationalization - all that is power depends on them."[527] Technique is, basically, about the enhancement of power and those who possess technical knowledge have it.[528]

More precisely, this is achieved by technocrats employing knowledge, practices, and a language separating them from ordinary people, with a very clear barrier existing between exclusive and general practices, as exemplified in the difference between programming and using a computer.[529] Computer science is, then, not counterbalancing the technical system, but reinforcing it. Computerisation will not create a new society and bring various forms of urban, industrial, capitalist, political, and technical organisation, due to the "solidity and rigor of the technical system."[530]

Yet, since scientists themselves have to engage in a technological discourse when explaining their science to non-scientist, through which they become the object of expectations, too. Indeed, following the discourse, "they end up as its slaves and have

---

[524]Ibid., 104.
[525]Ibid., 24.
[526]Ibid., 26.
[527]Ibid., 27.
[528]Ibid., 25.
[529]Ibid., 26.
[530]Ibid., 113.

to follow the common path of progress."[531] In the name of this progress, then, the whole framework of society has to undergo transformations in terms of economy, intellectuality, and communications. The adaptation of everything is needed.[532] This process is not criticised by intellectuals or churches, both of which are afraid of seeming out of touch. Therefore, they want to engage in technisation.[533] "Whether we take the World Council of Churches or the papacy, they have become the privileged agent of technological enthusiasm. They are in a panic lest they should be thought to be behind the times, obscurantist, out of things."[534] The scientists themselves remain 'untouchables', however, due to the "omnipotence of sacrosant science."[535]

In terms of modern societies, then, Ellul sees the emergence of a Great Design that goes along with science and technology; attempting to shape humanity in accordance with technological requirements, of which four play a particularly significant role: people have to work well and reliably; the individual should not be bothered about collective matters and mainly consume, using it's spending power, because consumption is an absolute duty; and information should be accepted as it is presented, e.g. in the media: "In sum, the four requirements of the great design, which are well on the way to being met, will make it impossible for people to have an individual view of their own lives or of the reality of the world in which they live. Amid all the extolling of the world of communication and information, the great choice which is being made is that of ignorance. Such is the great design."[536]

As a solution to this undesirable state, Ellul suggest to introduce a very strict rule of responsibility. This entails that politicians, administrators, and technicians must be held personally responsible for what they decide and do. He concedes that in the 19th century, it was necessary to have freedom from responsibility in order to ensure the independence of political decision and the anonymity of public functioning, but this measures of protection are now counterproductive.[537]

---

[531] Ibid., 196.
[532] Ibid., 273.
[533] Ibid., 395.
[534] Ibid., 396.
[535] Ibid., 408.
[536] Ibid., 406.
[537] Ibid., 300.

2.4. Conclusion

Technology is in Ellul's view not so much problematic in itself, but in the dominant function it acquires in relation to the other domains of society. Ironically, this function results from initiatives of originally individual agents who through their interaction bring forth a structure superseding the self-determination of the individual. Thus, technology turns into much more than a mere reflection or discourse on techniques, since it evolves into the key force of a societal development that no longer is determinable. While all those not involved in technology have to adapt to its extensive requirements of normalisation, even the agents of technique themselves are objects of societal expectations and the inherent drive of alleged progress. Since technology is regarded as pivotal to the economic growth and internal stability of society, it is excluded from any critical assessment, notwithstanding its unpredictable effects on the fabric of nature and culture.

Ellul's gloomy conclusion is, then, that unavoidably, the technical development will lead to the destruction of the very society that fosters it. Certainly, European societies have been undergoing changes in terms of their systems of belief, values, politics and economy during the past decades. The emergence of heterogeneous societies, the reduced impact of the churches, the persistence of unemployment in spite of economic recovery, and the lack of strong collective ideals have among other factors drastically altered Europe yet again. One should bear in mind, however, that while societal stability might seem desirable, European history displays the vanity of such hope. Moreover, by the sheer fact of their acting, humans change themselves and their environment, and, consequently, society with it. Let me briefly summarise, then, how Ellul's findings pertain to our present context:

Firstly, research in general is seen as an expression of human potentiality. In the field of biotechnology, this freedom is also interpreted as a fundamental condition for exploring the micro-organic level in human and non-human life alike. While freedom without repression is the hallmark of modern self-understanding, the wish to penetrate all aspects of life, to control bio-chemical mechanisms at large, and to substantially enhance nature in its variety turns biotechnology into the epitome of this human self.

Secondly, this alleged enterprise of human freedom has revealed itself as an instrument of enslavement. For it seems, indeed, almost impossible to even imagine contemporary European societies without biotechnology. It has become the object of fascination, which means the submission of the individual to the technological system.

This submission damages the very notion of agency, turning the individual human being into the object of vague interests, influences and developments, i.e. the interests forming biotechnology, a process of inhibition rather than empowerment. As we have seen earlier, there is a strange tendency to dissolve the individual human agent at the institutional level, for which he allegedly forms the very basis.

Thirdly, as a technical system, biotechnology requires agreement on political and economic organisation, a dependent social structure and little violent opposition. The necessity of European concerted actions of harmonisation and funding clearly illustrate that biotechnology in fact is far more an instrument of society than an independent entity, even though the public primarily may perceive it in latter terms and freely submit e.g. to the requirements of technical standardisation.

Fourthly, techniques are fundamentally ambivalent, for they assert certainty and control, while they actually seldom can achieve goals with great success. It is, e.g., striking that the failure rate in developing new medication for a variety of reasons remains very high, while the demand for better products increase under the impression that medical success indeed is achievable. According to Ellul, these patterns could be broken by a true assessment of all the advantages and all disadvantages of a technique, also those that surpass the monetary level. Yet this seems at least for now unthinkable, as long as e.g. negative results will not be published.

Fifthly, the alleged rationality of technique, which also goes for the apparently rational attitude towards the use and enhancement of nature, may be much more a means of justification than meets the eye. In spite of the rather limited possibilities current biotechnology offers, the fascination with this field is increasing, taking on the role of the most important factor for societal progress, and thus becoming an unavoidable feature. To the extent that biotechnology is assigned such a salvatory potential, the language used for it often carries religious overtones.

Sixthly, this development is facilitated by the bluff presenting the discourse on techniques, namely "technology", as an expression of actual processes, pretending that words signify reality. By asserting this new reality, the state can legitimise its own power with science and technique rather than religiously or democratically. Science and the state become co-dependent, and there is little doubt that the emphasis on biotechnology prioritises public efforts e.g. in terms of funding and education. While biotechnology promises to deliver the goods demanded by the state, such as public health, economic growth and improved food production, it also binds public and private resources so that the whole framework of society has to undergo transformations in order to adapt.

Seventhly, this technical progress implies that certain humans are considered un-exploitable, because they do not have the mental or physical capacity to comply with the standards of constant and high performances. Biotechnology seems to accelerate this process with the emergence of the possibilities for genetic testing and e.g. pre-implantation selection. The difficulties of defining the humanity of a being clearly and the increasing demand on using e.g. embryonic stem cells indicate a shift in the view of human life that until now has proven rather difficult to assess commonly.

Finally, what Ellul suggests in order to solve this undesirable state, namely a very strict rule of responsibility, has also been launched currently, demonstrating the appropriateness of including his thoughts into our present reflections. Anders Nordgren, e.g., proposes four elements that should guide responsible scientists in one field of biotechnology, namely human genetics. Conceived as a form of 'imaginative casuistry', these elements are the moral imagination of envisioning different ethically relevant consequences, the learning from history, the participation in a broad public dialogue and the integration of ethical reflection with scientific practice.[538] Especially the latter point has gained acceptance at the European level, as we shall see in the following sub-chapter.

In sum, Ellul's analysis of technology in general has provided with helpful insights into the specific character of biotechnology. What originally seemed limited to technological development as such, has through biotechnology gained momentum in terms of impact. Current societies are, obviously, adapting even more to the requirements of the biotechnological development than to former technologies, in particular because the distance between subject and object of research seems to diminish so drastically. There is, after all, reasons why there are increasing demands to include ethical reflections into the process of research and development, indicating desires to reapproach a sector of research that somehow is considered distant. As long as biotechnology merely concerns non-human organisms, it might even be seen as the full realisation of the modern endeavour. With the human agent turning itself into the very object of biotechnological design, however, this apparently strong subject suddenly loses the stability it seemed to possess. Biotechnology does presuppose an agency similar to that of technology in general, namely of a strong agent instigating changes in objectified nature, and it is marked by the same irony of resulting in a

---

[538]Nordgren (2001), 84.

stratification of the allegedly autonomous subject, turning it into anything but the master of matter. At the same time, the scope of its techniques and the possible impact on the understanding of our humanity, our agency and our relationship to the non-human environment, all of which surpass traditional technology by the envisioned levels of penetration and fundamental change, call for politics that may appropriately conduct the further development of this field, a suggestion of which I shall now present.

3. THE POLITICS OF CONTEMPORARY BIOTECHNOLOGY

Similar to other forms of technology, biotechnology has in the course of our investigation appeared as an almost independent entity beyond immediate control. Thus, Ellul's understanding of technology can be cautiously applied to current biotechnology, different from other forms of technology in its less stable connection with the concept of modernity. After all, it has also shown 'anachronistic' features. On one hand, it is envisioned as a comprehensive answer to a set of needs identified in European societies: the need to enhance the biotechnical development, especially in order to compete on the global market; the need to establish favourable economic and political conditions for this development; the need to ensure public health, in particular by the continuous production of new and better medicine; and the need to foster research in areas that might be interesting for biotechnological production. This view on biotechnology places significant trust in the human ability to understand and perform, an optimism comparable with that of the industrial age, where first coal and steel and later nuclear energy provided utopian resorts. Of course, both of these technological fields did propel societal progress, although at a certain price, and abandoning them have proven to be an exhaustive process. On the other hand, the abandonment of the great designs characterising modern societies and the difficulties of defining common grounds in new multicultural settings make biotechnology seem oddly out of tune.

Furthermore, while formed in and by societal developments, biotechnology is also regarded as a locomotive of such development and, thus, somehow understood as an external force, seemingly occurring by own virtue rather than actively being implemented. Emerging within the boundaries of society, it nevertheless remains somehow alien.[539] It is not really clear, then, *who* wants biotechnology to happen the way it does. Still, it comes into being, because a number of individuals choose to act on the assumption that what they do will bring about the changes they anticipate. Biotechnology is, thus, more than a name, it is an institution that offers a setting within which the reflections of the individual are formed and forming.

Freely adapting to particular discourses and systems, which in turn facilitate the techniques forming biotechnology, the individual is dominating and submissive in terms of its speaking, thinking and doing. Of course, there are good arguments for regarding biotechnology as an enterprise dependent upon human freedom, since referring to the freedom of research only makes sense, if one assumes that humans are free and entitled to express this freedom through their deeds. Such an understanding of freedom includes a theory of will according to which human understanding can be translated into acts complying with it: Accordingly, the will is an instrument of reason forming the realm within which biotechnology is conceived and planned. Also in this sense, biotechnology expresses a modern view of human agency.

At the same time, the complexity of the individual initiatives undertaken seems to make it far more difficult to regulate biotechnology than e.g. traffic. In part, this might be caused by the lack of clear authorities able to define and enforce common standards and practices. As we have seen, there is e.g. agreement on defending the values of free research and of the integrity of the individual in Europe, but none on how this human individual can be defined e.g. in terms of its coming into existence and the further bearings of this for using material from the early stages of human development.

Deeply rooted in a particular view on the human being, biotechnology is nevertheless a strangely imprecise entity with clearly measurable results but far less clear mechanisms of what actually produces them. When exploring the precise nature of the name 'biotechnology', we saw that the interaction taking place in biotechnology is less coordinated than in an organisation, because the latter tends to have some clear structure of goal, command and execution.

While biotechnology in certain respects bears resemblance to an organisation, then, it differs from it in that the *institutional* adaptation and conformity emerge in spite of the absence of a precisely defined power. Hence, an inherent dialectic marks

---

[539]A tentative analogy for this relation could be the way in which the individual human being experiences its conscience or any form of self-reflection. Fully aware that thoughts occur in the mind of the thinker, they may nevertheless be experienced as not completely identical with ourselves.

biotechnology: it is evolving with a fundamental trust in the strong agent characterising modern technology, while the lack of a clearly defined centre of power in terms of its execution bears clear postmodern traits. Other modern societal schemes, such as education or hygiene, at some point had such centres, typically located in the will of the rulers in charge[540], by whose command they would permeate society, i.e. the minds of the individuals.[541] Biotechnology, however, seems to evade this mechanism. Indeed, the need for regulation in the field clearly indicates that it is not as such necessarily expressing the will of government or of the political class in society.

This causes the ambiguity of biotechnology also reflected by the politics designed to regulate it, e.g. on the level of the COE, oscillating between attempts to limit and to stimulate the development and application of its techniques. The lack of a singular political strategy results from the mixed feelings with which biotechnology is observed in the public sphere.

Whenever fascination is intertwined with fear, human actions tend to lack determination, because their goal remains opaque. It is not surprising, then, that the politics formed by individuals who may not be fully convinced of the benefits or dangers of biotechnology will lead to regulations that sometimes appear as puzzling.[542]

Moreover, the complexity of the field, with its fusion of diverse interests and use of often highly specialised techniques, almost prohibit taking a clear stance quickly.

---

[540] I realise that this is a simplification of matters, of course, for this will is in itself formed by a variety of societal impulses. Even the absolute monarch is still influence by the people he wishes to govern. Yet, there is little doubt that at least in the European societies before WW I, the monarch or the government possessed the means to execute policies at will. In a way, European fascism was a hyperbolic expression of this tendency, a rather brute caricature of such absolutist tendencies.

[541] There is a fundamental difference between these two examples, of course: with human freedom is rooted in human reason, education can be understood as human right to develop the faculties it possesses. Knowledge would in this regard be an integrative part of human emancipation from the slavery of ignorance. Hygiene is rather a matter of convention, by which certain forms of human life seem preferable to others. Granted, hygiene is also part of the modern project of public health, but there is no essential need to launder clothes to the extent it is done today. A stained suit does not necessarily inflict on our integrity, but it may affect our social standing. Both schemes are only possible, however, if the individual supports it. If I chose to rebuke learning as well as hygiene, both would fail to manifest themselves. Note, then, that the cooperation of the individual, whether by force or by free choice, is pivotal to any societal development.

[542] For example, when the German government decided not to allow for the creation of human embryos for stem cell research, yet accepting the import of spare embryos from Israel, the political signal was far from clear.

Above, we have discussed reasons for regarding biotechnology as beneficial or dangerous. Simply put, biotechnology is seen as the facilitator of societal progress, because it may contribute to the development of e.g. more resistant crops and more apt medical treatment in general.

Concomitantly, the human individual is seen as endangered by the use of human material, the possible involvement of re-enumeration and the increasing difficulty to commonly define human nature precisely, in particular with regards to the time of its beginning. In order to prevent such dangers, there have been diverse efforts to define at least minimal standards of biotechnological research and development, e.g. in the COE convention on biomedicine. Certain practices have been banned legally, e.g. human cloning, and others have been subjected to demands of transparency, as with the labelling of genetically modified food products.

These examples show how politics is employed as a regulatory instrument for relating biotechnology to society. Biotechnology is in this sense somehow recognised as a distinct entity that develops on its own terms, presenting products to a society suddenly realising what has become possible. It is hardly surprising, then, that this would result in a state of chock and rejection, as so clearly visible in the aftermath of that famous sheep Dolly.

To my mind, the fundamental problem is the lack of permeation and transparency, i.e. the free and open exchange of knowledge, attitudes and beliefs. In modern society, such an exchange occurred, too, but on far more common grounds and, therefore, also less explicitly. Standardisation and stratification were hallmarks of these societies, resting on the foundations and settings of Jacobine or Christian traditions.

In postmodern societies, such common grounds seem lost and establishing them torments the European public, for the transformation of allegedly former homogeneous into now heterogeneous societies has given raise to fears and aggressions, e.g. xenophobia, once considered cured. It has proven cumbersome, then, to locate the sources of public morality. Therefore, there are new initiatives to define the values on which European societies rest and the principles employed to safeguard them, e.g. in the charter of the EU.

In 2002, the European Commission launched with its sixth Community framework programme for research, technological development and demonstration activities a wider strategy seeking to develop life sciences and biotechnology in harmony with ethical values and societal goals, partly in order to strengthen European competitiveness.[543] The public attention and debate in the field is seen as a positive sign of civic responsibility and involvement. Furthermore, there are strong indications of a broad public openness in Europe towards entering "into complex weighting of benefits against disadvantages, guided by fundamental values."[544]

Hence, the action plan proposed by the commission and adopted by the European Parliament and the Council of Ministers contains measures to promote the resource base as well as its regulation, involving various stakeholders in biotechnology.[545] Clearly, the Commission wishes to integrate ethical reflections into the research and development process itself in order to bridge the gap between societal expectations and the technological development, which includes ethical reviews of biotechnological research proposals for Commission funding.[546]

The strategy is, thus, to combine excellence in research with comprehensive societal scrutiny, simply because "the regulatory oversight applied to the development (...) is the expression of societal choices."[547] In this view, biotechnology is merely an instrument of society, albeit of great importance on the European and global scale. It is, thus, somewhat demystified in relation to society, losing its aura of an independent concept existing apart from individual actions and reduced to a name for practices on the application of which society should agree. In a nutshell, then, the Commission presents an "initiative for a coherent, collaborative and sustained effort."[548]

The various COE instruments on biotechnology are other examples of such attempts to (re-)establish common grounds. In these deliberations, human freedom emerges as a central value and the principle of integrity as a means of defending this freedom over against initiatives that might endanger it. Biotechnology is not just happening in a value-free zone, then. On the contrary, the interests forming this field are mere expressions of underlying values that are pursued by the individual participant. There are in other words reasons why a researcher wishes to perform certain experiments and why a company might find his or her work interesting to support. The integration of ethics as a discipline of reflection and evaluation into this

---

[543]Commission of the European Communities: COM (2002) 27, 19.

[544]Ibid.

[545]The action plan is found within the communication of the Commission reprinted in the appendix.

[546]Commission of the European Communities: COM (2002) 27, action 14, 40.

[547]Commission of the European Communities: COM (2002) 27, 21.

[548]Commission of the European Communities: COM (2002) 27, 30.

process may help to clarify these reasons and, thus, facilitate a clear communication about them. Most importantly, the more the values and norms of society are included into this process, the more society will be willing to accept its outcomes. In doing so, biotechnology might lose some of its enigma and glory. It might just evolve into a field similar to other forms of craftsmanship, not substantially differing from producing steel or building railways. If focussed on the production of specific products, such as medication, rather than on vague strategies for enhancing nature, it may become a far more integrated part of society.

Whenever biotechnology is regarded as the sole solution to societal problems or envisioned as a great scheme of overcoming the inequities of e.g. the frail human being, it is likely to fail in the realisation of such goals as well as in the public perception. Instead, it would be helpful to refocus biotechnology on the production of what it can achieve, coordinating its techniques rather than seeing it as the solution itself. At that point, biotechnology can be discussed far less emotionally and politics will be able to regulate this field like any other on which society, its pressure groups and representatives seek to set their mark. At that point, biotechnology might have become contemporary rather anachronistic, less a riddle than a field of practices employed in order to create products. Of course, there still is a fundamental difference between producing buckets and treating humans, and there are reasons to accept that human life in its various stages and forms may not be turned into mere objects of interests. Therefore, it is paramount to foster a broad deliberation on the values of society and the ways in which we understand e.g. human existence and its boundaries. It makes sense to enable the individual citizen to take part in such deliberations and a democratic understanding of society will include such a quest for transparency quite naturally. For as long as there is at the least the impression of a separation between biotechnology as a societal subsystem and society at large, such communication is doomed to strand. In my opinion, the strategy proposed by the European Commission is in this respect a very important step towards this fundamental change in the conception of biotechnology. Together with the legal framework designed by the Council of Europe, structures now emerge within which biotechnology can be fostered and regulated in accordance with broader deliberations on its ethical presuppositions and societal impact.

What I propagate is a weak form of biotechnology, then, employed with respect for its own limitations as well as those of the individual human being on whose agency it relies. This does not entail banning biotechnology as such, but subjecting it to societal scrutiny in terms of its ethical implications. In a way, it is merely a set of techniques that we can use if we choose to. By avoiding the entrapment of regarding it as necessary, we might gain the freedom to carefully select among the possibilities it offers.

APPENDIX

European Convention on Human Rights and Biomedicine ETS 164 (Full text) with chart of signatures, ratifications and reservations.

Additional Protocol to the Convention on Human Rights and Biomedicine on the Prohibition of Cloning Human Beings ETS 168 (Full text) with chart of signatures, ratifications and reservations.

Additional Protocol to the Convention on Human Rights and Biomedicine on Transplantation of Organs and Tissues of Human Origin. ETS 186. (Full text) with chart of signatures, ratifications and reservations.

COUNCIL   CONSEIL
OF EUROPE   DE L'EUROPE

European Treaty Series - No. 164

CONVENTION FOR THE PROTECTION OF HUMAN RIGHTS AND DIGNITY OF THE HUMAN BEING WITH REGARD TO THE APPLICATION OF BIOLOGY AND MEDICINE: CONVENTION ON HUMAN RIGHTS BIOMEDICINE

Oviedo, 4.IV.1997

Preamble

The member States of the Council of Europe, the other States and the European Community, signatories hereto,

Bearing in mind the Universal Declaration of Human Rights proclaimed by the General Assembly of the United Nations on 10 December 1948;

Bearing in mind the Convention for the Protection of Human Rights and Fundamental Freedoms of 4 November 1950;

Bearing in mind the European Social Charter of 18 October 1961;

Bearing in mind the International Covenant on Civil and Political Rights and the International Covenant on Economic, Social and Cultural Rights of 16 December 1966;

Bearing in mind the Convention for the Protection of Individuals with regard to Automatic Processing of Personal Data of 28 January 1981;

Bearing also in mind the Convention on the Rights of the Child of 20 November 1989;

Considering that the aim of the Council of Europe is the achievement of a greater unity between its members and that one of the methods by which that aim is to be pursued is the maintenance and further realisation of human rights and fundamental freedoms;

Conscious of the accelerating developments in biology and medicine;

Convinced of the need to respect the human being both as an individual and as a member of the human species and recognising the importance of ensuring the dignity of the human being;

Conscious that the misuse of biology and medicine may lead to acts endangering human dignity;

Affirming that progress in biology and medicine should be used for the benefit of present and future generations;

Stressing the need for international co-operation so that all humanity may enjoy the benefits of biology and medicine;
Recognising the importance of promoting a public debate on the questions posed by the application of biology and medicine and the responses to be given thereto;

Wishing to remind all members of society of their rights and responsibilities;

Taking account of the work of the Parliamentary Assembly in this field, including Recommendation 1160 (1991) on the preparation of a convention on bioethics;

Resolving to take such measures as are necessary to safeguard human dignity and the fundamental rights and freedoms of the individual with regard to the application of biology and medicine,

Have agreed as follows:

Chapter I – General provisions

Article 1 – Purpose and object

Parties to this Convention shall protect the dignity and identity of all human beings and guarantee everyone, without discrimination, respect for their integrity and other rights and fundamental freedoms with regard to the application of biology and medicine.

Each Party shall take in its internal law the necessary measures to give effect to the provisions of this Convention.

Article 2 – Primacy of the human being

The interests and welfare of the human being shall prevail over the sole interest of society or science.

Article 3 – Equitable access to health care

Parties, taking into account health needs and available resources, shall take appropriate measures with a view to providing, within their jurisdiction, equitable access to health care of appropriate quality.

Article 4 – Professional standards

Any intervention in the health field, including research, must be carried out in accordance with relevant professional obligations and standards.

Chapter II – Consent

Article 5 – General rule

An intervention in the health field may only be carried out after the person concerned has given free and informed consent to it.
This person shall beforehand be given appropriate information as to the purpose and nature of the intervention as well as on its consequences and risks.

The person concerned may freely withdraw consent at any time.

Article 6 – Protection of persons not able to consent

1   Subject to Articles 17 and 20 below, an intervention may only be carried out on a person who does not have the capacity to consent, for his or her direct benefit.

2   Where, according to law, a minor does not have the capacity to consent to an intervention, the intervention may only be carried out with the authorisation of his or her representative or an authority or a person or body provided for by law.

    The opinion of the minor shall be taken into consideration as an increasingly determining factor in proportion to his or her age and degree of maturity.

3   Where, according to law, an adult does not have the capacity to consent to an intervention because of a mental disability, a disease or for similar reasons, the intervention may only be carried out with the authorisation of his or her representative or an authority or a person or body provided for by law.

The individual concerned shall as far as possible take part in the authorisation procedure.

4   The representative, the authority, the person or the body mentioned in paragraphs 2 and 3 above shall be given, under the same conditions, the information referred to in Article 5.
5   The authorisation referred to in paragraphs 2 and 3 above may be withdrawn at any time in the best interests of the person concerned.

Article 7 – Protection of persons who have a mental disorder

Subject to protective conditions prescribed by law, including supervisory, control and appeal procedures, a person who has a mental disorder of a serious nature may be subjected, without his or her consent, to an intervention aimed at treating his or her mental disorder only where, without such treatment, serious harm is likely to result to his or her health.

Article 8 – Emergency situation

When because of an emergency situation the appropriate consent cannot be obtained, any medically necessary intervention may be carried out immediately for the benefit of the health of the individual concerned.

Article 9 – Previously expressed wishes

The previously expressed wishes relating to a medical intervention by a patient who is not, at the time of the intervention, in a state to express his or her wishes shall be taken into account.

Chapter III – Private life and right to information

Article 10 – Private life and right to information

1   Everyone has the right to respect for private life in relation to information about his or her health.

2   Everyone is entitled to know any information collected about his or her health. However, the wishes of individuals not to be so informed shall be observed.

3   In exceptional cases, restrictions may be placed by law on the exercise of the rights contained in paragraph 2 in the interests of the patient.

Chapter IV –   Human genome

Article 11 – Non-discrimination

Any form of discrimination against a person on grounds of his or her genetic heritage is prohibited.

Article 12 – Predictive genetic tests

Tests which are predictive of genetic diseases or which serve either to identify the subject as a carrier of a gene responsible for a disease or to detect a genetic predisposition or susceptibility to a disease may be performed only for health purposes or for scientific research linked to health purposes, and subject to appropriate genetic counselling.

Article 13 – Interventions on the human genome

An intervention seeking to modify the human genome may only be undertaken for preventive, diagnostic or therapeutic purposes and only if its aim is not to introduce any modification in the genome of any descendants.

Article 14 – Non-selection of sex

The use of techniques of medically assisted procreation shall not be allowed for the purpose of choosing a future child's sex, except where serious hereditary sex-related disease is to be avoided.

Chapter V – Scientific research

Article 15 – General rule

Scientific research in the field of biology and medicine shall be carried out freely, subject to the provisions of this Convention and the other legal provisions ensuring the protection of the human being.

Article 16 – Protection of persons undergoing research

Research on a person may only be undertaken if all the following conditions are met:

i   there is no alternative of comparable effectiveness to research on humans;

ii   the risks which may be incurred by that person are not disproportionate to the potential benefits of the research;

iii   the research project has been approved by the competent body after independent examination of its scientific merit, including assessment of the importance of the aim of the research, and multidisciplinary review of its ethical acceptability,

iv   the persons undergoing research have been informed of their rights and the safeguards prescribed by law for their protection;

v   the necessary consent as provided for under Article 5 has been given expressly, specifically and is documented. Such consent may be freely withdrawn at any time.

Article 17 –   Protection of persons not able to consent to research

1   Research on a person without the capacity to consent as stipulated in Article 5 may be undertaken only if all the following conditions are met:

  i   the conditions laid down in Article 16, sub-paragraphs i to iv, are fulfilled;

  ii   the results of the research have the potential to produce real and direct benefit to his or her health;

  iii research of comparable effectiveness cannot be carried out on individuals capable of giving consent;

  iv   the necessary authorisation provided for under Article 6 has been given specifically and in writing; and

  v   the person concerned does not object.

2   Exceptionally and under the protective conditions prescribed by law, where the research has not the potential to produce results of direct benefit to the health of the person concerned, such research may be authorised subject to the conditions laid down in paragraph 1, sub-paragraphs i, iii, iv and v above, and to the following additional conditions:

i   the research has the aim of contributing, through significant improvement in the scientific understanding of the individual's condition, disease or disorder, to the ultimate attainment of results capable of conferring benefit to the person concerned or to other persons in the same age category or afflicted with the same disease or disorder or having the same condition;

ii   the research entails only minimal risk and minimal burden for the individual concerned.

Article 18 – Research on embryos *in vitro*

1   Where the law allows research on embryos *in vitro*, it shall ensure adequate protection of the embryo.

2   The creation of human embryos for research purposes is prohibited.

Chapter VI – Organ and tissue removal from living donors for transplantation purposes

Article 19 – General rule

1   Removal of organs or tissue from a living person for transplantation purposes may be carried out solely for the therapeutic benefit of the recipient and where there is no suitable organ or tissue available from a deceased person and no other alternative therapeutic method of comparable effectiveness.

2   The necessary consent as provided for under Article 5 must have been given expressly and specifically either in written form or before an official body.

Article 20 – Protection of persons not able to consent to organ removal

1   No organ or tissue removal may be carried out on a person who does not have the capacity to consent under Article 5.

2   Exceptionally and under the protective conditions prescribed by law, the removal of regenerative tissue from a person who does not have the capacity to consent may be authorised provided the following conditions are met:

i   there is no compatible donor available who has the capacity to consent;

ii   the recipient is a brother or sister of the donor;

iii   the donation must have the potential to be life-saving for the recipient;

iv   the authorisation provided for under paragraphs 2 and 3 of Article 6 has been given specifically and in writing, in accordance with the law and with the approval of the competent body;

v   the potential donor concerned does not object.

Chapter VII – Prohibition of financial gain and disposal of a part of the human body

Article 21 – Prohibition of financial gain

The human body and its parts shall not, as such, give rise to financial gain.

Article 22 – Disposal of a removed part of the human body

When in the course of an intervention any part of a human body is removed, it may be stored and used for a purpose other than that for which it was removed, only if this is done in conformity with appropriate information and consent procedures.

Chapter VIII – Infringements of the provisions of the Convention

Article 23 – Infringement of the rights or principles

The Parties shall provide appropriate judicial protection to prevent or to put a stop to an unlawful infringement of the rights and principles set forth in this Convention at short notice.

Article 24 – Compensation for undue damage

The person who has suffered undue damage resulting from an intervention is entitled to fair compensation according to the conditions and procedures prescribed by law.

Article 25 – Sanctions

Parties shall provide for appropriate sanctions to be applied in the event of infringement of the provisions contained in this Convention.

Chapter IX – Relation between this Convention and other provisions

Article 26 – Restrictions on the exercise of the rights

1   No restrictions shall be placed on the exercise of the rights and protective provisions contained in this Convention other than such as are prescribed by law and are necessary in a democratic society in the interest of public safety, for the prevention of

crime, for the protection of public health or for the protection of the rights and freedoms of others.

2   The restrictions contemplated in the preceding paragraph may not be placed on Articles 11, 13, 14, 16, 17, 19, 20 and 21.

Article 27 – Wider protection

None of the provisions of this Convention shall be interpreted as limiting or otherwise affecting the possibility for a Party to grant a wider measure of protection with regard to the application of biology and medicine than is stipulated in this Convention.

Chapter X – Public debate

Article 28 – Public debate

Parties to this Convention shall see to it that the fundamental questions raised by the developments of biology and medicine are the subject of appropriate public discussion in the light, in particular, of relevant medical, social, economic, ethical and legal implications, and that their possible application is made the subject of appropriate consultation.

Chapter XI – Interpretation and follow-up of the Convention

Article 29 – Interpretation of the Convention

The European Court of Human Rights may give, without direct reference to any specific proceedings pending in a court, advisory opinions on legal questions concerning the interpretation of the present Convention at the request of:

–   the Government of a Party, after having informed the other Parties;

–   the Committee set up by Article 32, with membership restricted to the Representatives of the Parties to this Convention, by a decision adopted by a two-thirds majority of votes cast.

Article 30 – Reports on the application of the Convention

On receipt of a request from the Secretary General of the Council of Europe any Party shall furnish an explanation of the manner in which its internal law ensures the effective implementation of any of the provisions of the Convention.

Chapter XII – Protocols

Article 31 – Protocols

Protocols may be concluded in pursuance of Article 32, with a view to developing, in specific fields, the principles contained in this Convention.

The Protocols shall be open for signature by Signatories of the Convention. They shall be subject to ratification, acceptance or approval. A Signatory may not ratify, accept or approve Protocols without previously or simultaneously ratifying accepting or approving the Convention.

Chapter XIII – Amendments to the Convention

Article 32 – Amendments to the Convention

1   The tasks assigned to "the Committee" in the present article and in Article 29 shall be carried out by the Steering

Committee on Bioethics (CDBI), or by any other committee designated to do so by the Committee of Ministers.

2   Without prejudice to the specific provisions of Article 29, each member State of the Council of Europe, as well as each Party to the present Convention which is not a member of the Council of Europe, may be represented and have one vote in the Committee when the Committee carries out the tasks assigned to it by the present Convention.

3   Any State referred to in Article 33 or invited to accede to the Convention in accordance with the provisions of Article 34 which is not Party to this Convention may be represented on the Committee by an observer. If the European Community is not a Party it may be represented on the Committee by an observer.

4   In order to monitor scientific developments, the present Convention shall be examined within the Committee no later than five years from its entry into force and thereafter at such intervals as the Committee may determine.

5   Any proposal for an amendment to this Convention, and any proposal for a Protocol or for an amendment to a Protocol, presented by a Party, the Committee or the Committee of Ministers shall be communicated to the Secretary General of the Council of Europe and forwarded by him to the member States of the Council of Europe, to the European Community, to any Signatory, to any Party, to any State invited to sign this Convention in accordance with the provisions of Article 33 and to any State invited to accede to it in accordance with the provisions of Article 34.

6   The Committee shall examine the proposal not earlier than two

months after it has been forwarded by the Secretary General in accordance with paragraph 5. The Committee shall submit the text adopted by a two-thirds majority of the votes cast to the Committee of Ministers for approval. After its approval, this text shall be forwarded to the Parties for ratification, acceptance or approval.

7   Any amendment shall enter into force, in respect of those Parties which have accepted it, on the first day of the month following the expiration of a period of one month after the date on which five Parties, including at least four member States of the Council of Europe, have informed the Secretary General that they have accepted it.

         In respect of any Party which subsequently accepts it, the amendment shall enter into force on the first day of the month following the expiration of a period of one month after the date on which that Party has informed the Secretary General of its acceptance.

Chapter XIV – Final clauses

Article 33 – Signature, ratification and entry into force

1   This Convention shall be open for signature by the member States of the Council of Europe, the non-member States which have participated in its elaboration and by the European Community.

2   This Convention is subject to ratification, acceptance or approval. Instruments of ratification, acceptance or approval shall be deposited with the Secretary General of the Council of Europe.

3   This Convention shall enter into force on the first day of the month following the expiration of a period of three months after the date on which five States, including at least four member States of the Council of Europe, have expressed their consent to be bound by the Convention in accordance with the provisions of paragraph 2 of the present article.

4   In respect of any Signatory which subsequently expresses its consent to be bound by it, the Convention shall enter into force on the first day of the month following the expiration of a period of three months after the date of the deposit of its instrument of ratification, acceptance or approval.

Article 34 – Non-member States

1   After the entry into force of this Convention, the Committee of Ministers of the Council of Europe may, after consultation of the Parties, invite any non-member State of the Council of Europe to accede to this Convention by a decision taken by the majority provided for in Article 20, paragraph d, of the Statute of the Council of Europe, and by the unanimous vote of the representatives of the Contracting States entitled to sit on the Committee of Ministers.

2   In respect of any acceding State, the Convention shall enter into force on the first day of the month following the expiration of a period of three months after the date of deposit of the instrument of accession with the Secretary General of the Council of Europe.

Article 35 – Territories

1   Any Signatory may, at the time of signature or when depositing its instrument of ratification, acceptance or

approval, specify the territory or territories to which this Convention shall apply. Any other State may formulate the same declaration when depositing its instrument of accession.

2   Any Party may, at any later date, by a declaration addressed to the Secretary General of the Council of Europe, extend the application of this Convention to any other territory specified in the declaration and for whose international relations it is responsible or on whose behalf it is authorised to give undertakings. In respect of such territory the Convention shall enter into force on the first day of the month following the expiration of a period of three months after the date of receipt of such declaration by the Secretary General.

3   Any declaration made under the two preceding paragraphs may, in respect of any territory specified in such declaration, be withdrawn by a notification addressed to the Secretary General. The withdrawal shall become effective on the first day of the month following the expiration of a period of three months after the date of receipt of such notification by the Secretary General.

Article 36 – Reservations

1   Any State and the European Community may, when signing this Convention or when depositing the instrument of ratification, acceptance, approval or accession, make a reservation in respect of any particular provision of the Convention to the extent that any law then in force in its territory is not in conformity with the provision. Reservations of a general character shall not be permitted under this article.

2   Any reservation made under this article shall contain a brief statement of the relevant law.

3   Any Party which extends the application of this Convention to a territory mentioned in the declaration referred to in Article 35, paragraph 2, may, in respect of the territory concerned, make a reservation in accordance with the provisions of the preceding paragraphs.

4   Any Party which has made the reservation mentioned in this article may withdraw it by means of a declaration addressed to the Secretary General of the Council of Europe. The withdrawal shall become effective on the first day of the month following the expiration of a period of one month after the date of its receipt by the Secretary General.

Article 37 – Denunciation

1   Any Party may at any time denounce this Convention by means of a notification addressed to the Secretary General of the Council of Europe.

2   Such denunciation shall become effective on the first day of the month following the expiration of a period of three months after the date of receipt of the notification by the Secretary General.

Article 38 – Notifications

The Secretary General of the Council of Europe shall notify the member States of the Council, the European Community, any Signatory, any Party and any other State which has been invited to accede to this Convention of:

a   any signature;

b   the deposit of any instrument of ratification, acceptance, approval or accession;

c any date of entry into force of this Convention in accordance with Articles 33 or 34;

d any amendment or Protocol adopted in accordance with Article 32, and the date on which such an amendment or Protocol enters into force;

e any declaration made under the provisions of Article 35;

f any reservation and withdrawal of reservation made in pursuance of the provisions of Article 36;

g any other act, notification or communication relating to this Convention.

In witness whereof the undersigned, being duly authorised thereto, have signed this Convention.

Done at Oviedo (Asturias), this 4th day of April 1997, in English and French, both texts being equally authentic, in a single copy which shall be deposited in the archives of the Council of Europe. The Secretary General of the Council of Europe shall transmit certified copies to each member State of the Council of Europe, to the European Community, to the non-member States which have participated in the elaboration of this Convention, and to any State invited to accede to this Convention.

List of declarations made with respect to treaty no. 164

Denmark:
Reservation contained in the instrument of ratification deposited on 10 August 1999 - Or. Eng. Period covered: 01/12/99 -

> Article 10, paragraph 2, concerning the right to information of registered persons
>
> According to this provision, all persons are entitled to know any information collected about his or her health. However, the wishes of individuals not to be so informed shall be observed.
>
> Danish legislation on registers provides that health information may be exempted from the registered person's right to information. Likewise, Section 10, paragraph 5, of the Public Administration Act (Act No. 572-19/12-1985) provides that material provided as a basis for the preparation of public statistics or scientific studies is not subject to access.

Declaration contained in the instrument of ratification deposited on 10 August 1999 - Or. Eng. Period covered: 01/12/99 -

> Article 20, paragraph 2, sub-paragraph ii, concerning the removal of regenerative tissue
>
> Under this provision, the removal of regenerative tissue, for example bone marrow, from a minor may be authorised in exceptional circumstances if the recipient is a brother or sister of the donor. However, regenerative tissue may not be transplanted from an under-age child to one of its parents. This limitation is not compatible with general practice in Denmark, under Section 13 of the act on medical examinations prior to the issue of a death certificate, post-mortem examinations, transplantation, etc (Act No.

402-13 / 6-1990) and in certain other countries, given that there are cases - albeit very rare - in which donation from a child to a parent cannot be replaced by any other realistic or equal treatment. In such cases, the donation has the potential to be life-saving for the recipient. The tissue in question will be regenerated more quickly in the child, and the actual surgical intervention is a minor one in which the only risk is the risk related to the anaesthesia. If this treatment option is excluded, the result may be that the child will lose his or her mother or father.

Declaration contained in the instrument of ratification deposited on 10 August 1999 - Or. Engl. Period covered: 01/12/99 -

In accordance with Article 35 of the Convention, Denmark declares that until further notice the Convention shall not apply to the Faroe Islands and Greenland.

Moldova:
Declaration contained in the instrument of ratification deposited on 26 November 2002- Or. Engl./Mol. Period covered: 01/03/03 -

According to Article 35 of the Convention, the Republic of Moldova declares that it will apply the provisions of the Convention only on the territory controlled by the Government of the Republic of Moldova until the full establishment of the territorial integrity of the Republic ol Moldova.

Turkey:
Declaration transmitted by a letter from the Deputy Permanent Representative of Turkey, dated 17 November 1997, registered at the Secretariat General on 18 November 1997 - Or. Engl.

The Government of the Republic of Turkey has, by Decree No. 97 / 9766 dated 7 August 1997, confirmed the signature of the Convention, done ad referendum on 4 April 1997 by Mr Riza TÜRMEN, Ambassador Extraordinary and Plenipotentiary, Permanent Representative of Turkey to the Council of Europe.

Declaration transmitted by a letter from the Deputy Permanent Representative of Turkey, dated 17 November 1997, registered at the Secretariat General on 18 November 1997 - Or. Engl.

The Republic of Turkey, in accordance with Article 36 of the Convention, reserves the right not to apply the provision of Article 20, paragraph 2, of the Convention, which authorises, under certain conditions, the removal of regenerative tissue from a person who does not have the capacity to consent, for the reason that this provision does not conform with the prohibition provided in Article 5 of the Law No. 2238 on Organ and tissue Removal, Preservation and Transplantation.

European Treaty Series - No. 168

ADDITIONAL PROTOCOL TO THE CONVENTION FOR THE PROTECTION OF HUMAN RIGHTS AND DIGNITY OF THE HUMAN BEING WITH REGARD TO THE APPLICATION OF BIOLOGY AND MEDICINE, ON THE PROHIBITION OF CLONING HUMAN BEINGS

Paris, 12.I.1998

Preamble

The member States of the Council of Europe, the other States and the European Community Signatories to this Additional Protocol to the Convention for the Protection of Human Rights and Dignity of the Human Being with regard to the Application of Biology and Medicine,

Noting scientific developments in the field of mammal cloning, particularly through embryo splitting and nuclear transfer;

Mindful of the progress that some cloning techniques themselves may bring to scientific knowledge and its medical application;

Considering that the cloning of human beings may become a technical possibility;

Having noted that embryo splitting may occur naturally and sometimes result in the birth of genetically identical twins;

Considering however that the instrumentalisation of human beings through the deliberate creation of genetically identical human beings is contrary to human dignity and thus constitutes a misuse of biology and medicine;

Considering also the serious difficulties of a medical, psychological and social nature that such a deliberate biomedical practice might imply for all the individuals involved;

Considering the purpose of the Convention on Human Rights and Biomedicine, in particular the principle mentioned in Article 1 aiming to protect the dignity and identity of all human beings,

Have agreed as follows:

Article 1

1   Any intervention seeking to create a human being genetically
        identical to another human being, whether living or dead, is
        prohibited.

2   For the purpose of this article, the term human being
        "genetically identical" to another human being means a human
        being sharing with another the same nuclear gene set.

Article 2

No derogation from the provisions of this Protocol shall be made
under Article 26, paragraph 1, of the Convention.

Article 3

As between the Parties, the provisions of Articles 1 and 2 of this
Protocol shall be regarded as additional articles to the Convention and
all the provisions of the Convention shall apply accordingly.

Article 4

This Protocol shall be open for signature by Signatories to the
Convention. It is subject to ratification, acceptance or approval. A
Signatory may not ratify, accept or approve this Protocol unless it has
previously or simultaneously ratified, accepted or approved the
Convention. Instruments of ratification, acceptance or approval shall
be deposited with the Secretary General of the Council of Europe.

Article 5

1   This Protocol shall enter into force on the first day of the month
        following the expiration of a period of three months after the

date on which five States, including at least four member States
of the Council of Europe, have expressed their consent to be
bound by the Protocol in accordance with the provisions of
Article 4.

2   In respect of any Signatory which subsequently expresses its
         consent to be bound by it, the Protocol shall enter into force on
         the first day of the month following the expiration of a period
         of three months after the date of the deposit of the instrument
         of ratification, acceptance or approval.

Article 6

1   After the entry into force of this Protocol, any State which has
         acceded to the Convention may also accede to this Protocol.

2   Accession shall be effected by the deposit with the Secretary
         General of the Council of Europe of an instrument of accession
         which shall take effect on the first day of the month following
         the expiration of a period of three months after the date of its
         deposit.

Article 7

1   Any Party may at any time denounce this Protocol by means
         of a notification addressed to the Secretary General of the
         Council of Europe.

2   Such denunciation shall become effective on the first day of the
         month following the expiration of a period of three months
         after the date of receipt of such notification by the Secretary
         General.

Article 8

The Secretary General of the Council of Europe shall notify the member States of the Council of Europe, the European Community, any Signatory, any Party and any other State which has been invited to accede to the Convention of:

a any signature;

b the deposit of any instrument of ratification, acceptance, approval or accession;

c any date of entry into force of this Protocol in accordance with Articles 5 and 6;

d any other act, notification or communication relating to this Protocol.

In witness whereof the undersigned, being duly authorised thereto, have signed this Protocol.

Done at Paris, this twelfth day of January 1998, in English and in French, both texts being equally authentic, in a single copy which shall be deposited in the archives of the Council of Europe. The Secretary General of the Council of Europe shall transmit certified copies to each member State of the Council of Europe, to the non-member States which have participated in the elaboration of this Protocol, to any State invited to accede to the Convention and to the European Community.

List of declarations made with respect to treaty no. 168

Netherlands:
Declaration contained in a Note Verbale from the Permanent Representation of the Netherlands, dated 29 April 1998, handed to the Secretary General at the time of signature, on 4 May 1998 - Or. Engl.

> In relation to Article 1 of the Protocol, the Government of the Kingdom of the Netherlands declares that it interprets the term "human being" as referring exclusively to a human individual, i.e. a human being who has been born.

COUNCIL CONSEIL
OF EUROPE DE L'EUROPE

European Treaty Series - No. 186

ADDITIONAL PROTOCOL TO THE CONVENTION ON HUMAN RIGHTS AND BIOMEDICINE CONCERNING TRANSPLANTATION OF ORGANS AND TISSUES OF HUMAN ORIGIN

Strasbourg, 24.I.2002

Preamble

The member States of the Council of Europe, the other States and the European Community signatories to this Additional Protocol to the Convention for the Protection of Human Rights and Dignity of the Human Being with regard to the Application of Biology and Medicine (hereinafter referred to as "Convention on Human Rights and Biomedicine"),

Considering that the aim of the Council of Europe is the achievement of greater unity between its members and that one of the methods by which this aim is pursued is the maintenance and further realisation of human rights and fundamental freedoms;

Considering that the aim of the Convention on Human Rights and Biomedicine, as defined in Article 1, is to protect the dignity and identity of all human beings and guarantee everyone, without discrimination, respect for their integrity and other rights and fundamental freedoms with regard to the application of biology and medicine;

Considering that progress in medical science, in particular in the field of organ and tissue transplantation, contributes to saving lives or greatly improving their quality;

Considering that transplantation of organs and tissues is an established part of the health services offered to the population;

Considering that, in view of the shortage of organs and tissues, appropriate action should be taken to increase organ and tissue donation, in particular by informing the public of the importance of organ and tissue transplantation and by promoting European co-operation in this field;

Considering moreover the ethical, psychological and socio-cultural problems inherent in the transplantation of organs and tissues;

Considering that the misuse of organ and tissue transplantation may lead to acts endangering human life, well being or dignity;

Considering that organ and tissue transplantation should take place under conditions protecting the rights and freedoms of donors, potential donors and recipients of organs and tissues and that institutions must be instrumental in ensuring such conditions;

Recognising that, in facilitating the transplantation of organs and tissues in the interest of patients in Europe, there is a need to protect individual rights and freedoms and to prevent the commercialisation of parts of the human body involved in organ and tissue procurement, exchange and allocation activities;

Taking into account previous work of the Committee of Ministers and the Parliamentary Assembly of the Council of Europe in this field;

Resolving to take such measures as are necessary to safeguard human dignity and the rights and fundamental freedoms of the individual with regard to organ and tissue transplantation,

Have agreed as follows:

Chapter I – Object and scope

Article 1 – Object

Parties to this Protocol shall protect the dignity and identity of everyone and guarantee, without discrimination, respect for his or her integrity and other rights and fundamental freedoms with regard to transplantation of organs and tissues of human origin.

Article 2 - Scope and definitions

1   This Protocol applies to the transplantation of organs and tissues of human origin carried out for therapeutic purposes.

2   The provisions of this Protocol applicable to tissues shall apply also to cells, including haematopoietic stem cells.

3   The Protocol does not apply:

   a   to reproductive organs and tissue;
   b   to embryonic or foetal organs and tissues;
   c   to blood and blood derivatives.

4   For the purposes of this Protocol:

   –   the term "transplantation" covers the complete process of removal of an organ or tissue from one person and implantation of that organ or tissue into another person, including all procedures for preparation, preservation and storage;

   –   subject to the provisions of Article 20, the term "removal" refers to removal for the purposes of implantation.

Chapter II – General provisions

Article 3 – Transplantation system

Parties shall guarantee that a system exists to provide equitable access to transplantation services for patients.

Subject to the provisions of Chapter III, organs and, where appropriate, tissues shall be allocated only among patients on an official waiting list, in conformity with transparent, objective and duly justified rules according to medical criteria. The persons or bodies responsible for the allocation decision shall be designated within this framework.

In case of international organ exchange arrangements, the procedures must also ensure justified, effective distribution across the participating countries in a manner that takes into account the solidarity principle within each country.

The transplantation system shall ensure the collection and recording of the information required to ensure traceability of organs and tissues.

Article 4 – Professional standards

Any intervention in the field of organ or tissue transplantation must be carried out in accordance with relevant professional obligations and standards.

Article 5 – Information for the recipient

The recipient and, where appropriate, the person or body providing authorisation for the implantation shall beforehand be given appropriate information as to the purpose and nature of the implantation, its consequences and risks, as well as on the alternatives to the intervention.

Article 6 – Health and safety

All professionals involved in organ or tissue transplantation shall take all reasonable measures to minimise the risks of transmission of any disease to the recipient and to avoid any action which might affect the suitability of an organ or tissue for implantation.

Article 7 – Medical follow-up

Appropriate medical follow-up shall be offered to living donors and recipients after transplantation.

Article 8 – Information for health professionals and the public

Parties shall provide information for health professionals and for the public in general on the need for organs and tissues. They shall also provide information on the conditions relating to removal and implantation of organs and tissues, including matters relating to consent or authorisation, in particular with regard to removal from deceased persons.

Chapter III – Organ and tissue removal from living persons

Article 9 – General rule

Removal of organs or tissue from a living person may be carried out solely for the therapeutic benefit of the recipient and where there is no suitable organ or tissue available from a deceased person and no other alternative therapeutic method of comparable effectiveness.

Article 10 – Potential organ donors

Organ removal from a living donor may be carried out for the benefit of a recipient with whom the donor has a close personal relationship as defined by law, or, in the absence of such relationship, only under the conditions defined by law and with the approval of an appropriate independent body.

Article 11 – Evaluation of risks for the donor

Before organ or tissue removal, appropriate medical investigations and interventions shall be carried out to evaluate and reduce physical and psychological risks to the health of the donor.

The removal may not be carried out if there is a serious risk to the life or health of the donor.

Article 12 – Information for the donor

The donor and, where appropriate, the person or body providing authorisation according to Article 14, paragraph 2, of this Protocol, shall beforehand be given appropriate information as to the purpose and nature of the removal as well as on its consequences and risks.

They shall also be informed of the rights and the safeguards prescribed by law for the protection of the donor. In particular, they shall be informed of the right to have access to independent advice about such risks by a health professional having appropriate experience and who is not involved in the organ or tissue removal or subsequent transplantation procedures.

Article 13 – Consent of the living donor

Subject to Articles 14 and 15 of this Protocol, an organ or tissue may be removed from a living donor only after the person concerned has given free, informed and specific consent to it either in written form or before an official body.

The person concerned may freely withdraw consent at any time.

Article 14 – Protection of persons not able to consent to organ or tissue removal

1   No organ or tissue removal may be carried out on a person who does not have the capacity to consent under Article 13 of this Protocol.

2   Exceptionally, and under the protective conditions prescribed by law, the removal of regenerative tissue from a person who does not have the capacity to consent may be authorised provided the following conditions are met:

i   there is no compatible donor available who has the capacity to consent;

ii   the recipient is a brother or sister of the donor;

iii the donation has the potential to be life-saving for the recipient;

iv   the authorisation of his or her representative or an authority or a person or body provided for by law has been given specifically and in writing and with the approval of the competent body;

v   the potential donor concerned does not object.

Article 15 – Cell removal from a living donor

The law may provide that the provisions of Article 14, paragraph 2, indents ii and iii, shall not apply to cells insofar as it is established that their removal only implies minimal risk and minimal burden for the donor.

Chapter IV – Organ and tissue removal from deceased persons

Article 16 – Certification of death

Organs or tissues shall not be removed from the body of a deceased person unless that person has been certified dead in accordance with the law.

The doctors certifying the death of a person shall not be the same doctors who participate directly in removal of organs or tissues from the deceased person, or subsequent transplantation procedures, or having responsibilities for the care of potential organ or tissue recipients.

Article 17 – Consent and authorisation

Organs or tissues shall not be removed from the body of a deceased person unless consent or authorisation required by law has been obtained.

The removal shall not be carried out if the deceased person had objected to it.

Article 18 – Respect for the human body

During removal the human body must be treated with respect and all reasonable measures shall be taken to restore the appearance of the corpse.

Article 19 – Promotion of donation

Parties shall take all appropriate measures to promote the donation of organs and tissues.

Chapter V – Implantation of an organ or tissue removed for a purpose other than donation for implantation

Article 20 – Implantation of an organ or tissue removed for a purpose other than donation for implantation

1   When an organ or tissue is removed from a person for a purpose other than donation for implantation, it may only be implanted if the consequences and possible risks have been explained to that person and his or her informed consent, or appropriate authorisation in the case of a person not able to consent, has been obtained.

2   All the provisions of this Protocol apply to the situations referred to in paragraph 1, except for those in Chapter III and IV.

Chapter VI – Prohibition of financial gain

Article 21 – Prohibition of financial gain

1   The human body and its parts shall not, as such, give rise to financial gain or comparable advantage.

The aforementioned provision shall not prevent payments which do not constitute a financial gain or a comparable advantage, in particular:

–   compensation of living donors for loss of earnings and any other justifiable expenses caused by the removal or by the related medical examinations;

–   payment of a justifiable fee for legitimate medical or related technical services rendered in connection with transplantation;

- compensation in case of undue damage resulting from the removal of organs or tissues from living persons.

2 Advertising the need for, or availability of, organs or tissues, with a view to offering or seeking financial gain or comparable advantage, shall be prohibited.

Article 22 – Prohibition of organ and tissue trafficking

Organ and tissue trafficking shall be prohibited.

Chapter VII – Confidentiality

Article 23 – Confidentiality

1 All personal data relating to the person from whom organs or tissues have been removed and those relating to the recipient shall be considered to be confidential. Such data may only be collected, processed and communicated according to the rules relating to professional confidentiality and personal data protection.

2 The provisions of paragraph 1 shall be interpreted without prejudice to the provisions making possible, subject to appropriate safeguards, the collection, processing and communication of the necessary information about the person from whom organs or tissues have been removed or the recipient(s) of organs and tissues in so far as this is required for medical purposes, including traceability, as provided for in Article 3 of this Protocol.

Chapter VIII – Infringements of the provisions of the Protocol

Article 24 – Infringements of rights or principles

Parties shall provide appropriate judicial protection to prevent or to put a stop to an unlawful infringement of the rights and principles set forth in this Protocol at short notice.

Article 25 – Compensation for undue damage

The person who has suffered undue damage resulting from transplantation procedures is entitled to fair compensation according to the conditions and procedures prescribed by law.

Article 26 – Sanctions

Parties shall provide for appropriate sanctions to be applied in the event of infringement of the provisions contained in this Protocol.

Chapter IX – Co-operation between Parties

Article 27 – Co-operation between Parties

Parties shall take appropriate measures to ensure that there is efficient co-operation between them on organ and tissue transplantation, inter alia through information exchange.

In particular, they shall undertake appropriate measures to facilitate the rapid and safe transportation of organs and tissues to and from their territory.

Chapter X – Relation between this Protocol and the Convention, and re-examination of the Protocol

Article 28 – Relation between this Protocol and the Convention

As between the Parties, the provisions of Articles 1 to 27 of this Protocol shall be regarded as additional articles to the Convention on Human Rights and Biomedicine, and all the provisions of that Convention shall apply accordingly.

Article 29 – Re-examination of the Protocol

In order to monitor scientific developments, the present Protocol shall be examined within the Committee referred to in Article 32 of the Convention on Human Rights and Biomedicine no later than five years from the entry into force of this Protocol and thereafter at such intervals as the Committee may determine.

Chapter XI – Final clauses

Article 30 – Signature and ratification

This Protocol shall be open for signature by Signatories to the Convention. It is subject to ratification, acceptance or approval. A Signatory may not ratify, accept or approve this Protocol unless it has previously or simultaneously ratified, accepted or approved the Convention. Instruments of ratification, acceptance or approval shall be deposited with the Secretary General of the Council of Europe.

Article 31 – Entry into force

1   This Protocol shall enter into force on the first day of the month following the expiration of a period of three months after the date on which five States, including at least four member States

of the Council of Europe, have expressed their consent to be bound by the Protocol in accordance with the provisions of Article 30.

2	In respect of any Signatory which subsequently expresses its consent to be bound by it, the Protocol shall enter into force on the first day of the month following the expiration of a period of three months after the date of the deposit of the instrument of ratification, acceptance or approval.

Article 32 – Accession

1	After the entry into force of this Protocol, any State which has acceded to the Convention may also accede to this Protocol.

2	Accession shall be effected by the deposit with the Secretary General of the Council of Europe of an instrument of accession which shall take effect on the first day of the month following the expiration of a period of three months after the date of its deposit.

Article 33 – Denunciation

1	Any Party may at any time denounce this Protocol by means of a notification addressed to the Secretary General of the Council of Europe.

2	Such denunciation shall become effective on the first day of the month following the expiration of a period of three months after the date of receipt of such notification by the Secretary General.

Article 34 – Notification

The Secretary General of the Council of Europe shall notify the member States of the Council of Europe, the European Community, any Signatory, any Party and any other State which has been invited to accede to the Convention of:

a   any signature;

b   the deposit of any instrument of ratification, acceptance, approval or accession;

c   any date of entry into force of this Protocol in accordance with Articles 31 and 32;

d   any other act, notification or communication relating to this Protocol.

In witness whereof the undersigned, being duly authorised thereto, have signed this Protocol.

Done at Strasbourg, this 24th day of January 2002, in English and in French, both texts being equally authentic, in a single copy which shall be deposited in the archives of the Council of Europe. The Secretary General of the Council of Europe shall transmit certified copies to each member State of the Council of Europe, to the non-member States which have participated in the elaboration of this Protocol, to any State invited to accede to the Convention and to the European Community.

REFERENCES

Agar, Nicholas: Biocentrism and the Concept of Life, in: *Ethics* 108 (October 1997), 147-168.

Agich, George J: *Autonomy and Long-Term Care*. New York, NY et al: Oxford University Press, 1993.

Almond, Brenda: Ethics in Engineering, in: Philippe Goujon, Bertrand Hériard Dubreuil: *Technology and Ethics. A European Quest for Responsible Engineering*. Leuven: Peeters 2001, 31-44.

Augustine: *Confessions and Enchiridion*. Newly translated and edited by Albert C. Philadelphia: Outler Westminster Press 1955 (= The Library of Christian Classics Vol. 7).

Bains, William, and Chris Evans: *The business of biotechnology*, in: Colin Ratledge and Bjørn Kristiansen (eds.): *Basic Biotechnology*. Second edition. Cambridge: Cambridge University Press 2001, 255-279.

Beauchamp, Tom L. and James F. Childress: *Principles of Biomedical Ethics*. Fifth Edition. Oxford et al.: Oxford University Press, 2001.

Bergsma, Jurrit, and David C. Thomasma: *Autonomy And Clinical Medicine. Renewing the Health Professional Relation With The Patient*. Dordrecht et al.: Kluwer Academic Publishers, 2000. (= International library of ethics, law, and the new medicine).

Böhme, Gernot: On Human Nature, in: Armin Grunwald et al.: *On Human Nature. Anthropological, Biological, and Philosophical Foundations*. Berlin et al.: Springer, 2002, 3-14.

Boyle, Philip J., Edwin R. DuBose, Stephen J. Ellingson, David E. Guinn, and David B. McCurdy: *Organizational Ethics in Health Care. Principles, Cases, and Practical Solutions*. San Francisco: Jossey-Bass, 2001.

Bovens, Mark: *The Quest for Responsibility. Accountability and Citizenship in Complex Organisations*. Cambridge et al.: Cambridge University Press, 1998.

Breidbach, Olaf: Gestalt Recognition and Internal Representation - A Report from the Philosophical Laboratory, in: Armin Grunwald et al.: *On Human Nature. Anthropological, Biological, and Philosophical Foundations.* Berlin et al.: Springer, 2002, 81-94.

Bretell, Richard R.: *Modern Art, 1851-1929: Capitalism and Representation.* New York,NY: Oxford University Press, 1999. (=Oxford History of Arts).

Bretell, Richard R.: *Impression: Painting Quickly in France, 1860-1890.* Williamstown, MA: Sterling and Francine Clark Art Institute, 2000.

Brown, Jane: *The art and architecture of English Gardens.* London: Weidenfeld and Nicolson, 1989.

Buchanan, Allen and, Dan W. Brock, Norman Daniels, Daniel Wikler: *From Chance to Choice. Genetics and Justice.* Cambridge et al.: Cambridge University Press 2000.

Camus, Albert: *The Myth of Sisyphus and other Essays.* Translated by Justin O'Brien. New York, NY: Random House, 1955. (=Vintage Books 75).

Changeux, Jean-Pierre and Paul Ricoeur: *What Makes Us Think? A Neuroscientist and a Philosopher Argue about Ethics, Human Nature, and the Brain.* Transl. by M.B. DeBevoise. Princeton: Princeton University Press, 2000.

Chomsky, N.: *Rules and Representations.* New York, NY: Columbia University Press, 1980.

Critchley, Simon: The Prolegomena to Any Post-Deconstructive Subjectivity, in: Simon Critchley and Peter Dews (eds.): *Deconstructive Subjectivities.* New York, NY: State University of New York Press, 1996, 13-46.

Cole, R. David : The Genome and the Human Genome Project, in: Ted Peters: *Genetics. Issues of Social Justice.* Cleveland: Pilgrim Press, 1998 (= The Pilgrim Library of Ethics).

Collste, Göran: *Globalisation from an ethical perspective,* in: Philippe Goujon, Bertrand Hériard Dubreuil: *Technology and Ethics. A European Quest for Responsible Engineering.* Leuven: Peeters, 2001, 423-438.

Commission of the European Communities: *European competitiveness report 2001.* Luxembourg: Eur-Op, 2001.

Commission of the European Communities: *Life sciences and biotechnology - A strategy for Europe.* Communication COM (2002) 27 of 23 January 2002.

Cottingham, John: General introduction. The Meditations and Cartesian Philosophy, in: Descartes, R.: *Meditations on First Philosophy.* With Selections from the Objections and Replies. Translated and Edited by John Cottingham. With an introductory essay by Bernard Williams and a new introduction for this edition by John Cottingham. Revised edition. Cambridge et al.: Cambridge University Press, 2000, xviii-xlvi.

Council of Europe: Convention for the Protection of Human Rights and Dignity of the Human Being with regard to the Application of Biology and Medicine: Convention on Human Rights and Biomedicine. Oviedo, 4 April 1997. European Treaty Series 164.

Council of Europe Committee of Ministers: Resolution (78) 29 on harmonisation of legislation of member states relating to removal, grafting and transplantation of human substances, in: Council of Europe, Directorate of Legal Affairs: *Texts of the Council of Europe on bioethical matters.* CDBI/INF (98) 5. Strasbourg 1998, 46-50.

Council of Europe Committee of Ministers: Recommendation R (79) 5 of the Committee of Ministers to the member states concerning international exchange and transportation of human substances, in: Council of Europe, Directorate of Legal Affairs: *Texts of the Council of Europe on bioethical matters.* CDBI/INF (98) 5. Strasbourg 1998, 51-52.

Council of Europe Committee of Ministers: Recommmendation R (84)16 of the Committee of Ministers to member states concerning notification of work involving recombinant deoxyribonucleic acid, in: Council of Europe, Directorate of Legal Affairs: *Texts of the Council of Europe on bioethical matters.* CDBI/INF (98) 5. Strasbourg 1998, 57-59.

Council of Europe Committee of Ministers: Recommendation R (90) 3 of the Committee of Ministers to member states concerning medical research on human beings, in: Council of Europe, Directorate of Legal Affairs: *Texts of the Council of Europe on bioethical matters.* CDBI/INF (98) 5. Strasbourg 1998, 60-64.

Council of Europe Committee of Ministers: Recommendation R (90) 13 of the Committee of Ministers to member states on prenatal genetic screening, prenatal genetic diagnosis and associated genetic counselling, in: Council of Europe, Directorate of Legal Affairs: *Texts of the Council of Europe on bioethical matters.* CDBI/INF (98) 5. Strasbourg 1998, 65-68.

Council of Europe Committee of Ministers: Recommendation R (92) 1 of the Committee of Ministers to member states on the use of analysis of deoxyribonucleic acid (DNA) within the framework of the criminal justice system, in: Council of Europe, Directorate of Legal Affairs: *Texts of the Council of Europe on bioethical matters.* CDBI/INF (98) 5. Strasbourg 1998, 69-73.

Council of Europe Committee of Ministers: Recommendation R (92) 3 of the Committee of Ministers to member states on genetic testing and screening for health care purposes, in: Council of Europe, Directorate of Legal Affairs: *Texts of the Council of Europe on bioethical matters.* CDBI/INF (98) 5. Strasbourg 1998, 74-79.

Council of Europe Committee of Ministers: Recommendation R (93) 4 of the Committee of Ministers to member states concerning clinical trials involving the use of components and fractionated products derived from human blood or plasma, in: Council of Europe, Directorate of Legal Affairs: *Texts of the Council of Europe on bioethical matters.* CDBI/INF (98) 5. Strasbourg 1998, 80-85.

Council of Europe Committee of Ministers: Recommendation R (94) 1 of the Committee of Ministers to member states on human tissue banks, in: Council of Europe, Directorate of Legal Affairs: *Texts of the Council of Europe on bioethical matters.* CDBI/INF (98) 5. Strasbourg 1998, 86-88.

Council of Europe Committee of Ministers: Recommendation R (97) 5 of the Committee of Ministers to member states on the protection of medical data, in: Council of Europe, Directorate of Legal Affairs: *Texts of the Council of Europe on bioethical matters.* CDBI/INF (98) 5. Strasbourg 1998, 89-99.

Council of Europe Committee of Ministers: Recommendation R (97) 15 of the Committee of Ministers to member states on xenotransplantation, in: Council of Europe, Directorate of Legal Affairs: *Texts of the Council of Europe on bioethical matters.* CDBI/INF (98) 5. Strasbourg 1998, 100.

Council of Europe, Directorate of Legal Affairs: *Explanatory Report to the Convention for the Protection of Human Rights and Dignity of the Human Being with regard to the Application of Biology and Medicine: Convention on Human Rights and Biomedicine.* Doc DIR/JUR (97) 1, Strasbourg January 1997.

Craske, Matthew: *Art in Europe 1700-1830: A History of the Visual Arts in an Era of Unprecedented Urban Economic Growth.* Oxford: Oxford University Press, 1997. (=Oxford History of Art).

Damas, Maurice: *L'archéologie industrielle en France.* Paris: Lafont, 1980. (=Les hommes et l'histoire).

Danermaerk, Berth, and Ingemar Elander: *Social rented housing in Europe: policy, tenure and design.* Delft: University Press Delft, 1994. (=Housing and Urban Policy Studies 9).

de Brabandere, Luc: Condamnés à l'oxymoron, in: *La Libre Belgique,* 6 June 2002, 15.

de Haan, Juriaan: The New Dutch Law on Euthanasia, in: *Medical Law Review* 10 (2002), 57-75.

de Haan, Juriaan: The Ethics of Euthanasia: Advocates' Perspectives, in: *Bioethics* 16 (2002), 154-172.

de Finance SJ, Joseph: *An Ethical Inquiry.* Translated and adapted by Michael O'Brien SJ. Roma: Editrice Pontificia Università Gregoriana 1991.

den Hartogh, Govert: Priorities in Collective Health Care Provision: Why the Search for Criteria Failed, in: G. den Hartogh (ed.): *The Good Life as a Public Good.* Dordrecht et al.: Kluwer Academic Publishers, 2000, 107-118. (= Library of Ethics and Applied Philosophy, ed. by G.A. den Hartogh, 6).

Dekkers, Wim: *The human body*, in: H.A.M.J. ten Have and B. Gordijn (eds.): *Bioethics in a European Perspective.* Dordrecht et al.: Kluwer Academic Publishers 2001, 115-139. (= David C. Thomasma et al.: International Library of Ethics, Law the New Medicine, Vol.8).

Deleuze, Gilles and Félix Guattari: *What is Philosophy?* Translated by Hugh Tomlinson and Graham Burchell. New York, NY: Columbia University Press, 1994.

Descartes, R.: *Meditations on First Philosophy*. With Selections from the Objections and Replies. Translated and Edited by John Cottingham. With an introductory essay by Bernard Williams and a new introduction for this edition by John Cottingham. Revised edition. Cambridge et al.: Cambridge University Press, 2000.

Dulbecco, Renato: *The Design of Life*. New Haven: Yale University Press, 1987.

Duncker, Hans-Rainer: The Biological Fundamentals, in: Armin Grunwald et al.: *On Human Nature. Anthropological, Biological, and Philosophical Foundations*. Berlin et al.: Springer, 2002, 53-72.

Dupré, Louis: *Passage to Modernity. An Essay in the Hermeneutics of Nature and Culture*. New Haven and London: Yale University Press, 1993.

Ebert, Wolfgang and Achim Bednortz: *Kathedralen der Arbeit: historische Industriearchitektur in Deutschland - Cathedrals of work: historical industrial architecture in Germany*. Tübingen: Wasmuth, 1996.

Ehrlich, Paul R.: *Human natures: genes, cultures, and the human prospect*. Washington DC/Covelo, CA: Island Press/Shearwater Books, 2000.

Elliot, John: Say Goodbye to Privacy, in: *The Times*. (16 June 2002), 14.

Ellul, Jacques: *The Technological Society*. Translated by Joachim Neugroschel. New York, NY: Continuum, 1980.

Ellul, Jacques: *The Technological Bluff*. Translated by Geoffrey W. Bromiley. Grand Rapids: Wm. B. Eerdmans, 1990.

Enge, Olaf Torsten; Carl Friedrich Schroër; Ailsa Mataj and Martin Classen: *Garden architecture in Europe 1450-1800: From the villa garden of the Italian Renaissance to the English landscape garden*. Köln: Taschen, 1990.

Epstein, Julia: *Altered conditions. Disease, medicine, and storytelling*. New York, NY: Routledge, 1995.

Esping-Andersen, Gøsta: *The Three Worlds of Welfare Capitalism*. Princeton, NJ: Princeton University Press, 1991.

European Commission: *General Report on the activities of the European Group on Ethics in Science and New Technologies To The European Commission 1998-2000.* Luxembourg: Office for Official Publications of the European Commission 2001.

Evans, Arthur B.: *Jules Verne Rediscovered.* Westport, CT: Greenwood Press, 1988.

Faragher, John Mack (ed.): *Rereading Frederick Jackson Turner: The Significance of the Frontier in American History, and other Essays .* New York: H. Holt, 1994.

Feenberg, Andrew, and Alastair Hannay: *Technology and the politics of knowledge.* Bloomington and Indianapolis: Idiana University Press, 1995. (= Indiana Series in the philosophy of technology).

Feenberg, Andrew: *Alternative Modernity. The Technical Turn in Philosophy and Social Theory.* University of California Press: Berkeley, Los Angeles, London 1995.

Feenberg, Andrew: Subversive Rationalization. Technology, Power, and Democracy, in: Andrew Feenberg and Alastair Hannay: *Technology and the politics of knowledge.* Bloomington and Indianapolis: Idiana University Press, 1995, 3-22. (= Indiana Series in the philosophy of technology).

Ferry, Luc (transl. by Carol Volk): *The New Ecological Order.* Chicago, IL: University of Chicago Press, 1995.

Fodor, J.A.: *The Modularity of Mind.* Cambridge, MA: MIT Press/Bradford Books, 1983.

Foucault, Michel: *Ethics. Subjectivity and Truth.* Edited by Paul Rabinow. Essential works of Foucault 1954-1984, Vol. I. New York, NY: The New Press, 1997.

Foucault, Michel: *Power.* Edited by James D. Faubion. Translated by Robert Hurley and others. Essential works of Foucault 1954-1984, Volume 3. London et al.: Allen Lane The Penguin Press, 2001.

Freccero, John: Autobiography and Narrative, in: Thomas C. Heller, Morton Sosna, David E. Wellbery (eds.): *Reconstructing Individualism. Autonomy, Individuality, and the Self in Western Thought.* Stanford: Stanford UP, 1986, 16-29.

Gay, Peter: *The Enlightenment. An interpretation. The Science of Freedom .* Reissue. New York/London: W.W. Norton & Co., 1996.

Gilbert, Scott F.: Genetic Determinism, in: Armin Grunwald et al.: *On Human Nature. Anthropological, Biological, and Philosophical Foundations*. Berlin et al.: Springer, 2002, 121-140.

Grunwald, Armin: Philosophy and the Concept of Technology, in: Armin Grunwald et al.: *On Human Nature. Anthropological, Biological, and Philosophical Foundations*. Berlin et al.: Springer, 2002, 179-194.

Gutmann, Mathias: Human Culture's Natures, in: Armin Grunwald et al.: *On Human Nature. Anthropological, Biological, and Philosophical Foundations*. Berlin et al: Springer, 2002, 195-240.

Hadrossek, P.: Vitoria, in: *Lexikon für Theologie und Kirche (LThK)*. 2nd ed., vol. 10. Freiburg: Herder, 1965, 823-825.

Harwood, Colin R., and Anil Wipat: *Genome management and analysis: prokaryotes*, in: Colin Ratledge and Bjørn Kristiansen (eds.): *Basic Biotechnology*. Second edition. Cambridge: Cambridge University Press, 2001, 59-93.

Heath, Gregory: *The Self and Communicative Theory*. Aldershot et al.: Ashgate, 2001. (= Avebury series in philosophy).

Helland, Dag E.: Genomic understanding of life, in: Nordic Council of Ministers: *Who Owns Our Genes. Proceedings of an International Conference October 1999, Tallinn, Estonia*. Organised by the Nordic Committee on Bioethics. Copenhagen 2000, 11-21.

Hellin, J.: Vazquez, in: *Dictionnaire de Théologie Catholique*. Contentant l'exposé des doctrines de la théologie catholique, leurs preuves et leur historie. Commencé sous la direction de A. Vacant et E. Mangenot, continué sous celle de Mgr É. Amann. Vol I, 2nd part: Trinité-Zwinglianisme. Paris: Letouzey et Ané, 1950, 2601-2610.

Hösle, Vittorio: *Philosophie der ökologischen Krise: Moskauer Vorträge*. München: Beck'sche Verlagsbuchhandlung 1991. (= Beck'sche Reihe 432)

Hogenhuis, C. and D. Koelega: Engineers' Tools for Inclusive Technological Development, in: Philippe Goujon and Bertrand Hériard Dubreuil: *Technology and Ethics. A European Quest for Responsible Engineering*. Leuven: Peeters 2001, 207-229.

Humphrey, N.K.:: *A History of the Mind: Evolution and the Birth of the Consciousness*. London: Chatto and Windus, 1992.

Huntington, Samuel P: The Lonely Superpower, in: *Foreign Affairs* 78 (No 2, March/April 1999), 35-49.

Janich, Peter: Between Natural Dispostion and Cultural Masterment of Life, in: Armin Grunwald et al.: *On Human Nature. Anthropological, Biological, and Philosophical Foundations.* Berlin et al.: Springer, 2002, 95-110.

John Paul II, papa: *The Splendor of Truth Shines. Encyclical Letter Veritatis Splendor.* Vatican City: Libreria Editrice Vaticana, 1993.

Jonsen, Albert R.: *The Birth of Bioethics.* New York/Oxford: Oxford University Press, 1998.

Kierkegaard, Søren: *Philosophical Fragments or A Fragment of Philosophy.* By Johannes Climacus. Responsible for Publication Søren Kierkegaard. Originally translated and introduced by David E. Swenson. New Introduction and Commentary by Niels Thulstrup. Translation revised and commentary translated by Howard V. Hong. 2nd edition, 4th printing. s.l..: Princeton University Press, 1967.

Klein, Kerwin Lee: *Frontiers of Historical Imagination: Narrating the European Conquest of Native America, 1890-1990.* Berkeley: University of California Press, 1997.

Kollek, Regine: *Technicalisation of human procreation and social living conditions*, in: H. Haker/D. Beyleveld: *The Ethics of Genetics in Human Procreation.* Ashgate: Aldershot etc. 2000, 139-162.

Latour, Bruno: *We have never been modern.* Trans. by Catherine Porter. Cambridge, MA: Harvard University Press, 1993.

LeGoff, Jacques: *La Civilisation de l'Occident Médiéval.* Paris: Edition Arthaud, 1964.

LeGoff, Jacques: *La Nouvelle Histoire.* Paris: Éditions Complexe 1988 (= Historiques 47).

Lenoir, Noëlle: Europe Confronts the Embryonic Stem Cell Research Challenge, in: *Science* 287(2000) , 1425-1427.

Levin, David Michael and George F. Solomon: The Discursive Formation of the Body in the History of Medicine, in: *Journal of Medicine and Philosophy* 15 (1990), 515-537.

Lowe, David A.: *Antibiotics*, in: Colin Ratledge and Bjørn Kristiansen (eds.): *Basic Biotechnology*. Second edition. Cambridge: Cambridge University Press, 2001, 349-375.

Luhmann, Niklas: The Individuality of the Individual: Historical Meanings and Contemporary Problems, in: Thomas C. Heller, Morton Sosna, David E. Wellbery (eds.): *Reconstructing Individualism. Autonomy, Individuality, and the Self in Western Thought*.Stanford: Stanford UP, 1986, 313-325.

MacIntyre, Alasdair: *Dependent rational animals*. London: Duckworth, 1999.

Mahoney, John: *The Making of Moral Theology. A Study of the Roman Catholic Tradition*. Oxford: Clarendon, 1989.

Melehy, Hassan: *Writing Cogito. Montaigne, Descartes, and the Institution of the Modern Subject*. Albany: State University of New York Press, 1997. (= SUNY series, the margins of literature, ed. by Mihai I. Spariosu).

Meulenbergs, Tom: *Corruptie, vuile handen en algemeen belang. En reflectie op de moraliteit van het politieke bedrijf.* Unpublished master thesis. Leuven 2000.

Meulenbergs, Tom and Paul Schotsmans: The Sanctity of Autonomy? Transcending the opposition between a quality of life and a sanctity of life ethic, in: *Bijdragen. International Journal in Philosophy and Theology* 62 (2001), 280-303.

Mill, J.St.: *Utilitarianism. On Liberty. Essay on Bentham, together with selected writings of Jeremy Bentham and John Austin*. Edited with an Introduction by Mary Warnock. 17th Impression. Glasgow: Fontana 1985.

Millgram, E.: Mill's Proof of the Principle of Utility, in: *Ethics* 110 (2000), 282-310.

Moore, G. E.: *Elements of Ethics*. Edited and with an introduction by T. Regan. Philadelphia: Temple University Press 1991.

More, Sir Thomas: *Utopia*. Translated by Ralph Robynson 1556. Edited with an Introduction by David Harris Sacks. Boston/New York: Bedford/St. Martin's, 1999. (= The Bedford Series in History and Culture).

Nielsen, Geert A. and the Danish Ministry of Food, Agriculture and Fisheries: *Flagship in a food chain*. Copenhagen: The Danish Ministry of Food, Agriculture and Fisheries, 2001.

Nordgren, Anders: *Responsible Genetics. The Moral Responsibility of Geneticists for the Consequences of Human Genetics Research*. Dordrecht et al.: Kluwer Academic Publishers, 2001. (=Philosophy and Medicine, ed. By H. Tristam Engelhardt Jr.. Vol. 70)

Nowell, Nanette, in: *Encyclopedia of Bioethics*. Revised edition by Warren T. Reich (editor in chief). Volume 1. New York, NY: Simon and Schuster Macmillan, 1995, 283.

Nys, Herman: Doelstellingen, Leidende Beginselen en Mogelijke Beperkingen, in: H. Nys (ed.): *De Conventie Mensenrechten en Biogeneeskunde van de Raad van Europa. Inhoud en gevolgen voor patiënten en hulpverleners*. Antwerpen/Groningen: Intersentia Rechtswetenschappen, 1998, 43-63.

Onfray, Michel: interview given in: *Victor. Le magazine polysensuel du journal Le Soir*, no 15 (12 April 2002), 6.

Opinion of the European Group on Ethics in Science and New Technologies to the European Commission: Ethical Aspects of Human Stem Cell Research and Use. No 15, 14 November 2000, in: European Commission : *General Report on the activities of the European Group on Ethics in Science and New Technologies To The European Commission 1998 - 2000*. Luxembourg: Office for Official Publications of the European Commission 2001, 121-135.

Oyama, Susan: The Nurturing of Natures, in: Armin Grunwald et al.: *On Human Nature. Anthropological, Biological, and Philosophical Foundations*. Berlin et al.: Springer, 2002, 163-170.

Parliamentary Assembly of the Council of Europe: Recommendation 934 (1982), in: Council of Europe, Directorate of Legal Affairs: *Texts of the Council of Europe on bioethical matters*. CDBI/INF (98) 5. Strasbourg 1998, 15-18.

Parliamentary Assembly of the Council of Europe: Recommendation 1046 (1986) on the use of human embryos and foetuses for diagnostic, therapeutic, scientific, industrial and commercial purposes, in: Council of Europe, Directorate of Legal Affairs: *Texts of the Council of Europe on bioethical matters*. CDBI/INF (98) 5. Strasbourg 1998, 19-23.

Parliamentary Assembly of the Council of Europe: Recommendation 1100 (1989) on the use of human embryos and foetuses in scientific research, in: Council of Europe, Directorate of Legal Affairs: *Texts of the Council of Europe on bioethical matters*. CDBI/INF (98) 5. Strasbourg 1998, 24-30.

Parliamentary Assembly of the Council of Europe: Recommendation 1160 (1991) on the preparation of a convention on bioethics, in: Council of Europe, Directorate of Legal Affairs: *Texts of the Council of Europe on bioethical matters.* CDBI/INF (98) 5. Strasbourg 1998, 33-34.

Parliamentary Assembly of the Council of Europe: Recommendation 1213 (1993) on developments in biotechnology and the consequences for agriculture, in: Council of Europe, Directorate of Legal Affairs: *Texts of the Council of Europe on bioethical matters.* CDBI/INF (98) 5. Strasbourg 1998, 35-38.

Parliamentary Assembly of the Council of Europe: Recommendation 1240 (1994) on the protection and patentability of material of human origin, in: Council of Europe, Directorate of Legal Affairs: *Texts of the Council of Europe on bioethical matters*. CDBI/INF (2001) 2. Strasbourg 2001, 37-39.

Parliamentary Assembly of the Council of Europe: Recommendation 1399 (1999) on xenotransplantation, in: Council of Europe, Directorate of Legal Affairs: *Texts of the Council of Europe on bioethical matters*. CDBI/INF (2001) 2. Strasbourg 2001, 45-46.

Parliamentary Assembly of the Council of Europe: Recommendation 1418 (1999) on the protection of the human rights and dignity of the terminally ill and the dying, in: Council of Europe, Directorate of Legal Affairs: *Texts of the Council of Europe on bioethical matters.* CDBI/INF (2001) 2. Strasbourg 2001, 47-50.

Parliamentary Assembly of the Council of Europe: Recommendation 1425 (1999) on biotechnology and intellectual property, in: Council of Europe, Directorate of Legal Affairs: *Texts of the Council of Europe on bioethical matters.* CDBI/INF (2001) 2. Strasbourg 2001, 51-52.

Parliamentary Assembly of the Council of Europe: Recommendation 1468 (2000) on biotechnologies, in: Council of Europe, Directorate of Legal Affairs: *Texts of the Council of Europe on bioethical matters.* CDBI/INF (2001) 2. Strasbourg 2001, 53-54.

Parliamentary Assembly of the Council of Europe: Recommendation 1512 (2001) on the protection of the human genome by the Council of Europe, in: Council of Europe, Directorate of Legal Affairs: *Texts of the Council of Europe on bioethical matters.* CDBI/INF (2001) 2. Strasbourg 2001, 55-57.

Patrão-Neves, M.: The identity of the person, in: R.K. Lie and P.T. Schotsmans (eds.): *Healthy Thoughts. European Perspectives on Health Care Ethics.* Leuven: Peeters, 2002, 49-67. (= European Ethics Network Core Materials for the Development of Courses in Professional Ethics.)

Pimentel, David, and Kelsey Hart: *Pesticide use: ethical, environmental, and public health implications*, in: Arthur W. Galston/Emily G. Shurr: *New Dimensions in Bioethics. Science, Ethics and the Formulation of Public Policy.* Dordrecht et al.: Kluwer Academic Publishers, 2001, 79-108.

Pinckaers OP, Servais: *The Sources of Christian Ethics.* Translated from the third edition by Sr. Mary Thomas Noble OP. Washington D.C.: The Catholic University of America Press, 1995.

Power, Anne: *Hovels to high rise: State housing in Europe since 1850.* London: Routledge, 1993.

Power, Anne: *Estates on the Edge. The social consequences of mass housing in Northern Europe.* Basingstoke: MacMillan, 1997.

Prak, Niels L., and Hugo Priemus (eds.): *Post-war public housing in trouble.* Papers presented at the congress 'post-war public housing in trouble' conference in Delft, the Netherlands, October 4-5, 1984. Delft: University Press Delft, 1985.

Ratledge, Colin and Bjørn Kristiansen (eds.): *Basic Biotechnology.* Second edition. Cambridge: Cambridge University Press, 2001.

Ratledge, Colin: *Biochemistry and physiology of growth and metabolism*, in: Colin Ratledge and Bjørn Kristiansen (eds.): *Basic Biotechnology.* Second edition. Cambridge: Cambridge University Press 2001, 17-44.

Rehmann-Sutter, Christoph: Genetics, Embodiment and Identity, in: Armin Grunwald et al.: *On Human Nature. Anthropological, Biological, and Philosophical Foundations.* Berlin et al.: Springer, 2002, 23-50.

Reuter, Lars: Human is what is born of a human: Personhood, Rationality, and An European Convention, in: *Journal of Medicine and Philosophy* 25 (2000), 181-194.

Ross, W. D. Ross: *Foundations of Ethics.* Oxford et al.: Oxford University Press, 1939.

Sartre, Jean-Paul: Being and Nothingness. An Essay on Phenomenological Ontology. Translated and with in an introduction by Hazel E. Barnes. New York: Philosophical Library, 1956.

Scharf, Hans-Wolfgang and Friedhelm Ernst: *Vom Fernschnellzug zum Intercity. Die Geschichte des deutschen Schienenschnellverkehrs.* Freiburg: Eisenbahn-Kurier Verlag, 1983.

Schienstock, Gerd: Towards a European information economy, in: *Poeisis and Praxis* 1 (2001): 47-65.

Schotsmans, Paul: The Belgian Euthanasia Debate: Some new developments since December 1999, in: *Bulletin of medical ethics / European association of centres of medical ethics (EACME) news*, (2000) 160, 4f.

Schotsmans, Paul and Bert Broeckaert: Debating Euthanasia in Belgium, Part Two, in: *Hastings Center Report* 29 (1999) 5, 47f.

Schwinger, Elke: *Angewandte Ethik. Naturrecht, Menschenrechte.*München, Wien: R. Oldenbourg, 2001. (= Lehr- und Handbücher der Politikwissenschaft).

17th Conference of European Ministers of Justice (Istanbul 5-7 June 1990): Resolution No. 3 on bioethics, in: Council of Europe, Directorate of Legal Affairs: *Texts of the Council of Europe on bioethical matters*. CDBI/INF (98) 5. Strasbourg 1998, 112.

Sloan, Phillip R.: From Natural Law to Evolutionary Ethics in Enlightenment French Natural History, in: Jane Maienschein and Michael Ruse: Biology and the Foundations of Ethics. Cambridge: Cambridge University Press 1999, 52-83. (=Cambridge Studies in Philosophy and Biology).

Smith, John E.: *Public perception of biotechnology*, in: Colin Ratledge and Bjørn Kristiansen (eds.): *Basic Biotechnology*. Second edition. Cambridge: Cambridge University Press, 2001, 3-16.

Spencer, Edward M., Ann E. Mills, Mary V. Rorty, and Patricia H. Werhane: *Organization Ethics in Health Care.* New York et al.: Oxford University Press, 2000.

Stratton-Lake, Philip (ed.): *Ethical Intuitionism: Re-evaluations.* Oxford et al.: Oxford University Press, 2002.

Tambuyzer, Erik: Long term effects of Biotechnology on Health Care Systems & Services, in: Leiner, G. et al.: *Health and Social Security. Creating a better Future for Health in Europe. Congress Report European Health Forum Gastein 1999.* Bad Gastein: EHFG, 2000, 192-197.

ten Have, Henk A.M.J: *Introduction: Bioethics and European Traditions*, in: H.A.M.J. ten Have and B. Gordijn (eds.): *Bioethics in a European Perspective.* Dordrecht et al.:: Kluwer Academic Publishers, 2001, 1-11. (= David C. Thomasma et al.: International Library of Ethics, Law and the New Medicine, Vol.8).

ten Have, Henk A.M.J.: Theoretical models and approaches to ethics, in: H.A.M.J. ten Have and B. Gordijn (eds.): *Bioethics in a European Perspective.* Dordrecht et al:: Kluwer Academic Publishers, 2001, 51-82. (= David C. Thomasma et al.: International Library of Ethics, Law and the New Medicine, Vol.8).

ten Have, Henk A.M.J.: Genetics and culture, in: H.A.M.J. ten Have and B. Gordijn (eds.): *Bioethics in a European Perspective.* Dordrecht et al:: Kluwer Academic Publishers, 2001, 351-368. (= David C. Thomasma et al.: International Library of Ethics, Law and the New Medicine, Vol.8).

Theofilatou, M.A.: *The emerging health agenda. The health policy of the European community.* Doctoral dissertation at the University of Maastricht. s.l.: 2000.

St. Thomas Aquin: *Summa Theologiæ 1a 2æ, 90-97, vol. 28: Law and Political Theory.* Ed. by Thomas Gilby OP. Cambridge: Blackfriars, 1964.

St. Thomas Aquin: *Summa Theologiæ 1a 2æ ? - 11, vol. 2: Existence and nature of God.* Ed. by Timothy McDermott OP. Cambridge: Blackfriars, 1966.

St. Thomas Aquinas: *Summa Theologiæ 1a2æ 6-17, vol. 17: Psychology of human acts.* Ed. by Thomas Gilby OP. Cambridge: Blackfriars, 1970.

St. Thomas Aquin: *Summa Theologiæ 1a 2æ, 68-70, vol. 24: The Gifts of the Spirit.* Ed. by Edward D. O'Connor OP. Cambridge: Blackfriars: 1974.

St. Thomas Aquinas: *On Evil.* Translated by Jean Oesterle. Notre Dame: University of Notre Dame Press 1995.

Thomasma, David C.; Thomasine Kimbrough-Kushner; Gerrit Kimsma; and Chris Ciesielski-Carlucci: Asking to die. Inside the Dutch debate about euthanasia. Dordrecht et al.: Kluwer Academic Publishers, 1998.

Torres, Juan Manuel: The importance of genetic services for the theory of health: A basis for an *integrating view* of health, in: *Medicine, Health Care and Philosophy* 5 (2002), 43-51.

Trouet, Caroline: Vergelijkende analyse van de artikelen van de ontwerpteksten (1994-1996) en van de verdragstekst, in: H. Nys (ed.): *De Conventie Mensenrechten en Biogeneeskunde van de Raad van Europa. Inhoud en gevolgen voor patiënten en hulpverleners*. Antwerpen/Groningen: Intersentia Rechtswetenschappen, 1998, 65-110.

Tsue, Meng: *The Works of Mencius*. Translated by James Legge. New York, NY: Dover, 1970.

Yang, Huanming, Director Human Genome Center and Secretary-General, Chinese Human Genome Project: Advances in Human Genome Research and Their Impact on the Progress of Humanity, in: Division of Human Sciences, Philosophy and the Ethics of Science and Technology of UNESCO: *Proceedings International Bioethics Committee of UNESCO* (IBC), Seventh Session. Volume II. November 2000, 63-75.

VandeVeer, Donald and Christine Pierce: *The environmental ethics and policy book: philosophy, ecology, economics*. Belmont, CA: Wadsworth, 1994.

Verstraeten, Johan: Oorlog en politiek in het denken van Carl von Clausewitz, in: Res Publica. Belgian Journal for Political Science, Vol. 27 (1985), 31-57.

Watt, Fiona M., and Brigid L.M. Hogan: Out of Eden: Stem Cells and Their Niches, in: *Science* 287 (2000), 1427-1430.

Weber, Max: *The Protestant Ethic and the Spirit of Capitalism*. 3rd edition, new translation and introduction by Stephen Kalberg. Los Angeles, CA: Blackwell Publishing, 2002.

Weijer, Charles: I Need a Placebo Like I Need a Hole in the Head, in: *Journal of Law, Medicine and Ethics* 30 (2002), 69-72.

Wendler, Dave: Informed consent, exploitation and research, in: *Bioethics* 14 (2000), 310-339.

Wils, Jean-Pierre: Autonomy and Recognition, in: Hille Haker and Deryck Beyleveld: *The Ethics of Genetics in Human Procreation*. Aldershot et al.: Ashgate, 2000, 101-124.

Zilgalvis, Peteris: The European Convention on Human Rights and Biomedicine. Its Past, Present and Future, in: Peteris V. Zilgalvis et al.: *Healthcare Law: The Impact of the Human Rights Act 1998*. London: Cavendish Publishing, 2002.

Zwart, Hub: *Comment: Transcending the either/or: Can bioethics be critical?*, in: H. Haker/D. Beyleveld: *The Ethics of Genetics in Human Procreation*. Aldershot et al.: Ashgate, 2000, 163-168.

INDEX

agriculture . . . . 11, 13, 14 ,19, 37, 40, 114, 119, 215, 217

AIDS 18

Alzheimer's disease 44

anthropocentrism 3, 96, 128, 130, 144

autonomy . . . 3, 13, 14, 16, 44, 58, 65, 67, 70, 71, 74, 95, 97-103, 105, 107, 108, 110,
123, 126, 206, 212, 215, 221

autopsy 31, 46

biology . 15, 20, 31, 36, 37, 42, 55, 61, 69, 75, 82, 83, 111, 122, 162-165, 169, 174,
185, 186, 192, 208, 210, 219

biomedicine . . 15, 21, 24, 27, 29, 31, 36, 37, 39, 42, 45, 46, 53-55, 57, 58, 61, 62, 71,
157, 161, 162, 186, 191, 192, 203, 208, 210, 222

brain 10, 73, 84, 86, 91-94, 109, 110, 207

case law 43

categorisation 5, 10, 79, 126

Christianity 96, 102, 106, 133

cloning 19-23, 34, 39, 45, 54, 61, 62, 64, 115, 118, 157, 185, 186

colonialism 5, 9

conscience 56, 107, 121, 132, 134, 155

consent 46, 48, 50-52, 57-63, 65, 66, 68, 71, 75, 166, 167, 170-173, 178, 184, 188,
196-200, 204, 221

consumption 64, 136, 148, 150

contingency 114, 129

contrat social 29, 108

control . 12, 13, 27, 33, 39, 45, 52, 64, 68, 78, 80, 81, 88, 89, 93, 95, 97, 110, 113,
123, 126, 131, 132, 134, 138, 139, 145, 151, 152, 154, 167, 183

Convention on Biological Diversity 40

cortex 10, 91, 92

Council of Europe's Steering Committee on Bioethics (CDBI) 27

Creutzfeldt-Jacobs Disease 18

Dedicated Biotechnology Firms (DBF) 17, 18

disease   . 13, 18, 21, 22, 32, 33, 37, 44, 47, 50-53, 56, 58, 59, 62, 64, 75, 76, 87, 88,
            117, 125, 166, 168, 169, 171, 196, 211

divine authority 103

Divine law (lex divina) 106

DNA 12, 13, 21, 31, 41, 46, 47, 51, 82, 83, 115, 116, 122, 209

dogma of biology 82, 83

ectogenesis 34

embryos, human 22-24, 32-34, 36, 39, 46, 53, 57, 60, 61, 66, 156, 171, 216

embryonic stem cells 22, 23, 64, 112

enlightenment 3, 9, 67, 97, 102, 103, 107, 108, 133, 212, 219

enzymes 18, 19, 82

Eternal law (lex æterna) 103, 104, 106, 107

ethics   16, 23, 24, 26, 27, 30, 49, 65, 101, 102, 120, 158, 206, 207, 210, 212, 214-
           216, 218-221

Euratom 26

EURO 27

European biotechnology 17, 40

European Coal and Steel Community (ECSC) 26

European Commission 11, 17, 18, 22, 23, 26-28, 40, 118, 157, 159

European Council 21, 28

European Convention for the Protection of Human Rights and Fundamental Freedoms
           (ECHR) 30, 55

European Court of Human Rights 27, 29, 43, 174

(European) Court of Justice 41

European group on ethics to the commission 23, 24, 27

European Parliament 27, 29, 158

European public 157

European Social Charter 55, 56, 163

European societies .... 151, 154, 156, 157 euthanasia 42, 43, 98

Food and Agriculture Organisation (FAO) 40

free will 79, 109

genes 8, 34, 37, 40, 41, 45, 51, 63, 69, 82-84, 87

genetic disorder 44, 50, 51

genetic hermeneutics 83, 85

genetic modification 19, 20

genetic screening 46, 49-51

genetic testing 20, 45, 51, 52, 153

globalisation 5-7, 144

grace 74, 103, 105

Great Design 151, 155

haemophiliacs 18

health care 11, 14, 18, 51, 52, 58, 60, 61, 64, 65, 99, 100, 102, 166

health laws 60

hepatitis C 18

history 2, 4, 9, 11, 13, 52, 86, 104, 115, 147, 152, 154

human act 3, 8, 14, 35, 67, 72-76, 83, 88, 97, 102, 105, 111, 118-120

human actions 4, 72, 156

human activity 1, 5, 76, 77, 80, 105, 111, 114

human agent 3, 4, 14, 15, 20, 64, 65, 68, 69, 76, 97, 105, 106, 108, 111, 113, 138,
        139, 142, 152, 153

human agency 3, 7, 10, 14, 64, 68, 72, 82, 104, 155

human behaviour 110

human beings 10, 11, 19, 21, 33, 38, 39, 48, 51, 52, 54, 61, 62, 67, 72, 81, 92, 101,
        109, 134, 161, 165, 185, 186, 192, 208

human blood 46, 52, 53, 209

human body 38, 47, 53, 62, 84, 91, 96, 125, 141, 172, 173, 193, 199, 200, 210

human condition 70, 109

human dignity 21, 34, 41, 42, 45, 48-50, 54, 61, 66, 164, 186, 193

human existence . . . . 32, 38, 71, 76, 84, 90, 95, 104, 138, 159 human freedom 35, 46, 47, 54, 73-76, 84, 85, 100, 107, 109, 123, 138, 139, 151, 155, 156, 158

human genome . . . . see genome, human

Human Genome Project 2, 45, 207, 221

human identity 10, 73, 82, 116

human insulin 18

human integrity 66

human life . 33, 34, 57, 61, 62, 64, 66, 71, 75, 98, 103, 112, 113, 129, 151, 153, 156, 159, 193

human mind 3, 94, 95, 107

human organ 47, 82, 153

human person 8, 37, 38, 48, 50, 51, 103

human personhood 8, 38, 48, 50, 103

human potentiality 112, 138, 151

human reason 65, 68, 83, 102, 104, 105, 126, 130, 156

human rights 21, 24, 26, 27, 29-37, 39, 42-45, 50, 53-57, 58, 61, 62, 66, 67, 161-163, 174, 185, 186, 192, 203, 208, 210, 217, 222

human self. 129, 131

human self-determination 110

human subject 3, 13, 38, 64, 70, 71, 74, 76, 96, 97, 102, 107, 113, 126, 131, 134, 135

human substances 28, 31, 46, 47, 208

human tissue 31, 53, 209

identity . 2, 5, 10, 33, 35, 39, 55, 57, 72, 73, 77, 82, 84, 90, 97, 116, 122, 126, 132, 169, 186, 192, 194, 218

immanence 130, 131, 134

individuality 4, 66, 67, 77, 127, 132, 133, 212, 215

industrial revolution 11, 79

information 6, 19, 29, 30, 51, 58, 60, 62, 80, 84, 86, 93, 94, 112, 119, 132, 134, 145, 146, 149, 150, 166-168, 173, 182, 195-197, 201, 202, 219

informed consent 46, 50, 51, 58, 65, 166, 200, 221

institution . . 18, 25-27, 54, 59, 66, 67, 71, 72, 82, 84, 95, 97, 117-119, 122-124, 126, 132, 134, 146, 152, 155, 193, 215

International Court of Justice 28, 29
international law 30, 44, 67, 68, 105, 106
In-vitro-fertilisation (IVF) 23
ius civile 105, 107
ius gentium 105, 106
ius inter gentes 106
justice 28-30, 36, 41, 46, 50, 51, 105, 106, 148, 207, 209, 219
language . . 44, 47, 58, 59, 69, 74, 85, 87, 89, 117, 118, 122, 126, 142-144, 149, 152
law 2, 13, 15, 20, 29, 30, 34, 35, 37, 38, 42-45, 50, 52, 54-63, 65-68, 83, 97, 98, 101, 103-108, 110, 120, 132, 149, 165-168, 170-173, 175, 179, 184, 197-199, 202, 206, 210, 219-222
mastery 78, 79, 82, 83, 131, 137-139
medicine 12-15, 20, 21, 31, 33, 36, 37, 42, 55, 61, 76, 86, 87, 98, 111, 113, 125, 154, 162-165, 169, 174, 185, 186, 192, 206, 208, 210, 211, 214, 216, 218, 220, 221
micro-organisms 10, 14, 19, 20
modernity 1-5, 7-9, 11, 68, 84, 86, 87, 90, 96, 101, 103, 106, 114, 125-127, 130, 138, 154, 211, 212
narratives 126
national law 52, 58, 61
natural law (lex naturalis) 30, 103-107, 110, 219
nature . 1, 3, 5, 12, 13, 15, 32, 38, 51, 54, 57, 59, 61-64, 68-72, 75, 81, 83, 84, 88, 90, 92, 95, 96, 101, 102, 104-107, 108, 111-114, 117, 121, 124, 127, 129, 133, 138, 139, 140, 145, 146, 151-153, 155, 157, 159, 166, 167, 186, 195, 197, 206, 207, 211, 213, 214, 216, 218, 220
Parkinson's disease 22
patent law 38, 48, 63
periodisation 3, 4, 8-10, 100
personhood 8, 38, 48, 50, 54, 57, 58, 65, 70, 103, 128, 218
pharmaceuticals 37, 141
politics 41, 59, 151, 154, 156, 157, 159, 212
positive law 54, 105, 107
postmodernity 8, 86, 125, 127

power 3, 5, 7, 29, 69, 76, 78-81, 89, 108, 117, 122, 123, 131, 132

progress 1, 6, 33-37, 39, 40, 45, 49, 51, 58, 62, 68, 71, 114, 115, 135

privacy 53, 55, 116, 118

psychology 7, 75, 96

prenatal genetic screening 50

rationality . 38, 47, 57-59, 67, 72-75, 81, 101-103, 108, 111, 113, 125, 126, 134, 143,
          152

rDNA 12, 13

regulation 18, 37, 44, 48, 54, 66-68, 76, 98, 106, 109, 111, 112, 121, 156, 158

RNA 13, 14, 16

Single Market 38

Sixth Framework Programme (FP6) 17, 26, 157

small and medium sized companies (SME) 26

social rights 56

society   . 2, 5, 7, 9, 41, 45, 49, 59, 60, 64, 66-68, 80, 81, 96, 98, 106, 109, 112, 113,
          119-121, 127, 132-136, 139, 141-144, 146, 147, 149-152, 154, 156-159, 164,
          165, 173, 211, 227

somatic cell nuclear transfer (SCNT) 21

standardisation 6, 7, 144, 152

state   3, 30, 40, 42, 44, 67, 101, 126, 148, 152, 176, 178-181, 188, 189, 204, 205,
          218

stem cells 22, 23, 64, 112, 153, 194, 221

subject   1, 3, 12, 13, 58, 64, 68-71, 74-77, 95-97, 102, 107, 113, 117, 119, 123, 126,
          127, 129-135, 153, 154, 168, 215

subjectivity 3, 85, 128, 207, 212

technique .. 12, 14, 16, 19-22, 25, 30-33, 36, 38, 41, 47, 48, 51, 54, 63, 67-72, 75-82,
          111, 116, 118, 119, 132, 135, 137-149, 151, 152, 154-156, 159, 169, 186

technocrats 149

technological bluff 135, 137, 211

technology1, 6, 10-12, 14, 15, 33, 35, 37, 39, 49, 66, 69, 79-82, 86, 111, 113-115, 124,
          135-139, 141, 142, 144, 145, 147, 149, 151, 153, 154, 156, 206, 207, 212,
          213, 221

terminally ill 31, 42-44, 217

theological virtue 104

transcendence 104, 130, 131, 134

transparency 157, 159

transplantation . . 22, 28, 31, 39, 46, 47, 53, 54, 62, 63, 161, 171, 182, 184, 191-197,
                          199, 200, 202, 208, 209, 217

treatment 18, 19, 24, 32, 42-44, 47, 52, 65, 88, 112, 122, 149, 157, 167, 183

United Nations Declaration of Human Rights (UNDHR) 30, 55

utilitarianism 8, 215

utopia 2, 13, 154, 215

vegetative acts 75

World Health Organisation (WHO) 39

World Intellectual Property Organisation 40

World Trade Organisation 40

World War I (WW I) 114, 156

World War II (WW II) 115